Harald Pöcher

The defence equipment production of European states

Harald Pöcher

The defence equipment production of European states

Once and now

ScienciaScripts

Imprint

Any brand names and product names mentioned in this book are subject to trademark, brand or patent protection and are trademarks or registered trademarks of their respective holders. The use of brand names, product names, common names, trade names, product descriptions etc. even without a particular marking in this work is in no way to be construed to mean that such names may be regarded as unrestricted in respect of trademark and brand protection legislation and could thus be used by anyone.

Cover image: www.ingimage.com

This book is a translation from the original published under ISBN 978-620-2-32342-0.

Publisher:
Sciencia Scripts
is a trademark of
International Book Market Service Ltd., member of OmniScriptum Publishing Group
17 Meldrum Street, Beau Bassin 71504, Mauritius
Printed at: see last page
ISBN: 978-620-0-99507-0

Foreword

The book is the result of almost five years of research on the defence industry in Europe. However, the author points out that the present publication is not a scientific work, but rather presents the subject matter in an easily readable form for the interested reader. He also points out that the topic is a living matter and that companies have to adapt to the market and change their company name over time. In plain language, the present work has a topicality in the year 2020, which may already be outdated in the next year or the year after next.

In the first two sections, the author provides information on the questions "What is armaments and the armaments economy", "what is meant by the term dual-use goods" and "what legal aspects are connected with the procurement of armaments by a nation state in Europe".

In a further section, the author examines the historical significance of the defence industry for Europe in an overview, and finally, in the fourth section, he analyses the individual nation states and their defence industries. In order to avoid making the work too extensive, the author has only presented the most important armament enterprises in the individual countries.

The present book builds on previous work of the author on this topic. In particular, the author has written an extensive book on armament in Europe and several essays on armament economies of individual states and published them in the military history journal Pallasch.

The author thanks the Akademikerverlag for its willingness to include the book in its publishing program and to translate it into several languages with its translation software.

Honorary University Professor (NKE) Dr. habil Harald Pöcher, Brigadier General

Content

I. Terms

Armor

The use of violence has a long tradition in human history. While in ancient times it was still a matter of "bare" livelihood security, this changed with the emergence of organized armed armies. At the latest since absolutism the rulers went over to maintaining standing armies, which caused the first large permanent mass demand for armaments. In the following, armament is to be understood as all physical and psychological measures that are necessary to establish military protection. The increase or decrease of armament or armed forces on a considerable scale is called armament or disarmament. Closely related to the term armament are the terms armament economy and armament industry. Both terms will be explained in the next chapter.

In principle, goods which serve to equip and furnish armed forces are classified as military equipment in accordance with the list of war material annexed to Article 296 of the EU Treaty, formerly Article 223 of the Rome Treaties of 1958, and as dual-use goods. The Annex to Art. 296 is in an unauthorised English translation (the original is in French - see the Annex; there is also an attempt to translate it into German):

„The provisions of Article 223 (1) (b) of the Treaty of Rome apply to arms munitions and war material listed hereunder, including nuclear arms:

> *1. Automatic and portable firearms, such as rifles, carbines, revolvers, pistols, submachine guns and machine guns, but excluding sporting guns, pistols, and other small arms of a calibre less than 7 mm.*

> *2. Artillery equipment and smoke, gas and flame-throwing such as*

(a)   guns [korrekt: cannons], *howitzers, mortars, artillery pieces, anti-tank weapons, rocket launchers, flame-throwers, recoilless guns.*

(b)   Military equipment for firing smoke or gas.

3.   Ammunition intended for the arms listed in paragraphs 1 and 2 above.

4.   Bombs, torpedoes, rockets and guided missiles:

(a)   bombs, torpedoes, grenades, including smoke grenades, smoke canisters, rockets, mines, guided missiles, depth charges [underwater grenades], *incendiary bombs.*

(b)   Equipment and devices for military purposes, specially designed for handling, priming, de-priming, detonating or detecting the weapons [korrekt: articles] *listed in paragraph (a) above.*

5.   Fire control equipment for military purposes:

(a)   Firing computers, infra-red aiming devices and other night aiming equipment;

(b)   Range-finders, position indicators, altimeters;

(c)   Electronic, gyroscopic, optical and acoustic observation devices;

(d)   Bomb sights and gun sights, periscopes for the arms [korrekt: articles] *mentioned in this list.*

6.   Tanks and vehicles specially designed for military purposes:

(a)   Tanks;

(b)   Military-type vehicles, armed or armoured, including amphibious vehicles;

(c)   Armoured trains;

(d)   Half-track military vehicles;

(e)   Military vehicles for the recovery of tanks;

(f) Specially designed trailers for the transportation of ammunition listed in paragraphs 3 and 4.

7. Toxic or radioactive agents:

(a) Biological or chemical toxic agents and radioactive agents designed, in time of war, to kill people and animals or to destroy crops;

(b) Military equipment for spreading, detecting and identifying the substances listed in paragraph (a) above.

(c) Protective equipment against the substances listed in paragraph (a) above.

8. Charges, explosives and liquid or solid propellants:

(a) Charges and liquid or solid propellants specially designed and manufactured for the weapons [korrekt: material] *listed in paragraphs 3, 4 and 7 above;*

(b) Military explosives;

(c) Incendiary and gelatine incendiary constituents for military purposes.

9. Warship and specialised equipment therefore [korrekt: and their specialist equipment]:

(a) Warships of all types;

(b) Equipment specially designed for laying, detecting and sweeping mines;

(c) Submarine nets.

10. Aircraft and aircraft equipment [korrekt: and their specialist equipment] *for*

military purposes.

11. Electronic equipment for military purposes.

12. Photographic equipment specially designed for military purposes.

13. Other equipment and stores [correct: material]:

(a)   Parachutes and parachuting equipment [correct: parachute fabric];

(b)   Bridging equipment specially designed for military purposes;

(c)   Electrically-operated projectors for military purposes.

14.   Specialised parts and items for the equipment mentioned in this list insofar as

they are of a military nature.

Machinery, equipment and gear designed exclusively for research, manufacture, testing and inspection of arms, munitions and devices mentioned in this list intended solely for military purposes."

Armaments Industry

If one determines the needs of the military, "the demand spectrum of the military ranges from ornamental plants for the officers' mess to electronic devices". This already makes it clear that the attempt to define exactly what is included under military equipment or to completely record everything that is included under military equipment can hardly lead to a satisfactory result. Without wanting to write a scientific treatise here, it is tacitly assumed here that in an economy all goods of military demand are produced or at least can be produced. In the following section, the armaments industry is to be understood as that part of the national economy which has the task of producing all kinds of military equipment in order to cover the equipment requirements of the armed forces in the most economical way possible. The term "defence industry" goes far beyond the term "defence industry" and therefore also includes those companies and enterprises which produce so-called dual-use goods and are not limited to the production of hard military equipment. We will discuss the classification of armaments in more detail in the next sub-chapter.

Dual-use goods

The term "dual-use items" means items, including computer programs and technology, which can be used for both civil and military purposes. It includes all goods that can be used both for non-explosive purposes and for any form of assistance in the production of nuclear weapons or other nuclear explosive devices.

In 1996 the <u>Wassenaar Arrangement for the</u> Control of Exports of Conventional Arms, Dual-Use Goods, Technologies and Software was established as an instrument of regional and international security. Controlled trade in dual-use goods is intended to prevent a destabilising effect of such products in crisis regions. The Wassenaar Secretariat is based in Vienna.

The EU harmonised Community export control by adopting export control measures. The existence of a Community export control system is a prerequisite for the free movement of dual-use items, technology and software within the Community. The EC Dual-Use Regulation No. 1334/2000 as amended by the Council of 22 June 2000, published in the Official Journal EC L159 of 30.06.2000, is the Community regime for the control of exports of dual-use items and technology. These goods, technologies and software controlled in exports are listed in the annexes of the EC Dual-Use Regulation .

II Legal aspects of defence procurement in the EU

Procurement as such is not a specific feature of the 20th century, since the Roman Empire already had regulations on how to proceed when building temples and public buildings.

Defence equipment accounts for a large share of public procurement expenditure in the EU. The combined defence budgets of all Member States amount to some €170 billion, of which more than €80 billion is spent on procurement in general and €30 billion specifically on the purchase of new military equipment. However, most defence contracts are not awarded in accordance with EU rules under the Article 296 derogation, but on the basis of national procurement rules, which vary considerably from country to country. This can limit market access for foreign suppliers, leading to additional costs and inefficiencies, which have a negative impact on the competitiveness of the European defence industry.

The extensive use of Article 296 in defence procurement is not compatible with the EC Treaty and the case law of the European Court of Justice, which clearly states that the exemption may only be applied in specific, clearly defined cases.

EU public procurement rules help to make the best use of taxpayers' money. All public contracts above a certain threshold must be put out to public tender in accordance with the principles of transparency, equal treatment and non-discrimination.

The Directive on defence procurement (Directive 2009/81/EC) sets out specific EU-wide rules for the procurement of arms, munitions and war material (and related works and services) intended for defence purposes. It also establishes the legal framework for the procurement of sensitive goods, works and services for security purposes. These provisions are adapted to the specificities of the defence sector, which is usually extremely complex and sensitive. The directive provides the legal framework for improving the transparency and openness of defence markets among Member States, while at the same time safeguarding the security interests of individual countries.

The directive came into force on 21 August 2009 and had to be transposed into the national law of the EU countries by 21 August 2011.

In principle, public procurement rules should guarantee the highest level of transparency and thus prevent any manipulation of contracting authorities. However, when one looks at the equipment of the armed forces in Europe, the large states, which still have an efficient arms industry, make ample use of the exemptions, or they deliberately disregard EU law in the procurement of military equipment.

III. A brief outline of Europe's armament history

Like the laws of nature, war is part of human history. If we look at the historical epochs according to the standards of war history, we can undoubtedly conclude that firstly, in antiquity and the Middle Ages, secondly, between 1500 and 1700 the scale of warfare - measured by the mere size of armies and fleets - increased significantly, and thirdly, warfare became even more sophisticated, with the result that the amount of production resources used for weapons and other equipment and for fortifications increased more rapidly than the number of soldiers. Nevertheless, the demand for armaments remained rather modest until the establishment of mass armies in the second half of the 19th century, so that research and development in the armaments sector had little impact on civilian products. It was only when the demand for armaments skyrocketed in the age of imperialism and the advantage on the battlefield could also be ensured by new weapons that research and development in the armaments sector was strongly promoted. Many products of that time had a decisive influence on the development of civil products. The First World War was only a foretaste of the inventive talent of the individual states in the next great world fire, which entered the history books as the Second World War. Let us begin with a brief foray through the millennia.

Armor in the ancient world

Antiquity is understood to be the period of antiquity from around 1200 BC to 600 AD. It thus includes the advanced civilisations of Egypt, Mesopotamia, Syria, Asia Minor, Greece and the Roman Empire. In the rest of the world, the first advanced civilizations can be found in East Asia - especially in China, India and in Central and South America. Thanks to the modern methods of archaeology, one gains more and more insight into the living conditions, the economy and the exercise of power of individual peoples and cultures during the early years before Christ. Much is still to be researched, but much will always remain in the dark of history, such as the more or less exact data on the population development of that period. One is largely dependent on estimates, which, however, are very imprecise. Likewise, there are no exact data on the strength of the armies during that period and on the production of armaments. As an economist and historian one is therefore dependent on the sparse reports from that time. On the basis of these eyewitness reports, the scientist will be able to formulate hypotheses and, after further detailed investigations and tests, draw the logical conclusions that will ultimately provide a picture of the reality of that time. Since the time of Christ's birth, there has been an OECD study in which estimates of world populations have been made. According to this, around 230 million people lived around the birth of Christ, the majority of them, namely around 175 million in Asia. In Europe, including Russia, about 30 million people lived. The most densely populated countries were Italy with 7 million, France with 5 million, Spain with 4.5 million and Germany and Russia with 3 million each. In the other countries, between 100,000 and 800,000 people lived. See in detail the figure "World population in figures from the birth of Christ to 2006".

Let us first look at the demand side for military equipment, the armed forces of that time. Due to the small population, the personnel strength of the armed forces was also small compared to modern times. From the time of King Sargon of Akkad, an inscription from around 2250 B.C. is preserved, which tells us about the strength of his army. The inscription reads: "Sargon, King of Kish, has won victory in thirty-four battles; the walls he destroyed are as far away as the coasts of the sea. God Enlil has saved Sargon, the king, from the hand of his rivals. Fifty-four thousand men feed daily at his court." A standing army of 54,000 warriors gave the great king a sure superiority over any local rival. But maintaining such a large army also required annual campaigns and the devastation of many fertile lands to provide food for the soldiers. The burden on the population was obviously immense. Sargon's army, when invading enemy territory, could be likened to an epidemic, which sweeps away a significant portion of the enemy people, but which also brings with it long-term immunity, especially through its passage.

A great power of the ancient world was the empire of the pharaohs on the Nile. An important part of the society of the Egyptian state was the army. Until the time of the New Empire, the Egyptian army was an army that was only called up when needed. The pharaoh usually kept a small cadre of standing troops at his court, but in an emergency this could only form the core of an army. The majority were then farmers and craftsmen, who were called together by the officials in the provinces. This ad hoc summoned army was entrusted to the command of a member of the royal house or a capable official, who dismissed it when it was no longer needed. Towards the end of the first interim period, the expulsion of the Hyksos required a stable military organization. The men who founded the eighteenth dynasty could no longer do without a professional army. This professional army developed much later than the priesthood. When it had consolidated, it used its power for the benefit of the pharaoh and not against him. A testimony of the strength of the Egyptian army is the battle of Kadesh. This battle was fought during the conflict between the Hittites and the Egyptians around 1300 BC. Pharaoh Ramses II wanted to secure his access to Syria by taking the city of Kadesh, but the Hittite king Muwatalli prevented him from doing so. The Hittites and their 13 allies had an army of 3,700 chariots and 40,000 foot soldiers, the Egyptians had 2,000 chariots and 20,000 foot soldiers. The Hittites already had the first iron cutting weapons in these battles.

During the Persian Wars, Xerxes is said to have gathered a large army in Asia Minor in 481 B.C. It is reported that 800,000 men on foot, 80,000 horsemen and a fleet of 1,200 ships. With this contingent he wanted to subjugate Greece. After the Greek sea victory at Salamis, Xerxes returned to Asia Minor. In Greece, he left behind only 300,000 men. In comparison, the Greek army was modest. Only 100,000 men fought under the command of Pausanias and Aristeidis against the Persians at Plateaus. The Greeks, especially the Athenians and Spartans, won against the Persians. The Persian encampment, with its untold booty, fell into the hands of the Greeks.

Another commander of the ancient world, Alexander the Great set out for Asia Minor at the beginning of spring 334 B.C. with an army of 30,000 footmen and 5,000 horsemen. His army was brought to Asia Minor by the Macedonian fleet, 160 three-rower ships at Sestos. In the course of the battles against the Persians, who mostly had a force of several 100,000 men and horsemen, Alexander increased his forces by recruiting soldiers in the defeated territories. It is reported that for the intended subjugation of India, Alexander set off towards the end of 327 with 120,000 men from Bactria via Alexandria on the Paropamiso to north-western India.

Probably the most important period in military history after Alexander the Great can be seen as the period of the rise and fall of the Roman Empire. The Roman Empire began with the foundation of Rome in 753 BC. The first king was Romulus. It existed until 476 AD. The last emperor was Romulus Augustulus. The military constitution was changed again and again over the centuries. The Roman armies were in the older times real citizen armies, in which they were formed by elevation and the proletarians were excluded. Around 107 B.C. the proletarians were admitted to the army for the first time by Marius.

In the Roman Empire, the armaments industry played no role in the overall economy, since the personnel strength of the armed forces was also small in relation to the population. At the beginning of the 1st millennium, 25 legions stood at the borders. Without auxiliary troops, that was about 125,000 men - about half of the recipients of support, who lived on public welfare in the capital alone. The total annual burden on the state budget for salaries was about 25 million denarii (200 per capita)-just one tenth of the amount distributed by Augustus as donations among the people of Rome. Converted to the head of the empire's population, the cost of the army was only a few cents per year. Today, even a small country like Austria has a far greater burden on its national budget in terms of expenditure on national defence than the Roman Empire was. A comparison with the world powers of today seems absurd in every respect.

The military successes of the Romans were not based on their weapons, these were not miracle weapons and differed not insignificantly from those of their opponents. Every legion had to provide for the weapons and tools and had to set up the necessary factories. In addition there were commodities, such as metal tableware and bronze objects. The factories and workshops operated by the legion were located in the Canabae legionis outside the camp. For example, the canabae legions of the city of Vindobona, today's Vienna, were located in the area of today's Michaeler-Platz. They were under the control of a specially appointed officer, the praefectus fabrum. The need for weapons and equipment was so great that in addition to the army workshops, the armatoria, civilian entrepreneurs also came to their business and established such fabricae armorum. Weapon and tool smiths were respected people, since the production of swords and lance tips or entire armour required great skill in handling iron and also required artistry.

The Roman emperors, who were also supreme commanders of the armed forces, also played a major role in supplying the army with weapons and goods. During the imperial period the economic power of the emperor was great. The emperors were the largest landowners and for military and economic reasons they reserved the right to dispose of a number of provinces. This made supplying the army a profitable business for the emperor as well. While the army was the mainstay of power in the Roman Empire, the fleet remained insignificant except for a few outstanding sea victories. Only the fleet at the border rivers Rhine and Danube gained importance while the Germanic invasions increased.

Warfare in ancient times was limited mainly by transport and provisioning. The supply of metals and weapons was also essential, but as a variable it was rarely decisive. There was therefore no real "industrial" aspect to warfare. Nevertheless, a number of significant changes in weapon systems can be discovered in history books, the result of sporadic technical discoveries and inventions that brought about a change in existing conditions of warfare and army organisation. A first change of this kind was brought about by the advent of bronze weapons. This began in the prehistory of advanced civilizations around 3500 B.C. in Mesopotamia. Then, several thousand years later, the next significant change in weapon systems occurred. It was the result of an efficient construction of chariots. This significantly improved mobility and firepower. The chariot became a superior means of warfare, which, mainly due to the invention of the spoke wheel, had a much smoother running than the chariots with wooden disc wheels. When the construction of the chariot was perfected, a good archer standing next to the charioteer was able to throw arrows at enemy foot soldiers, while he himself was relatively safe thanks to the rapid movement of the chariot. The chariots played an important role in the Orient, India and China from 1500 BC onwards, while in Europe the chariots apparently played a less important role. The chariots were expensive to produce because of the amount of work involved in their construction and operation, as food for the draught horses had to be available all year round. Societies dominated by chariots were therefore markedly aristocratic. A very small warrior class claimed the majority of agricultural surpluses, which were squeezed out of the farmers, for itself. Social conditions changed rapidly in the opposite direction when the next change in weapon systems occurred. With the discovery of iron and the possibility of producing cheaply usable weapons from iron, a radical democratization of warfare took place somewhere in eastern Asia Minor from 1,400 BC onwards. As iron ore deposits were widespread, the metal became enormously cheaper and the charcoal needed for the smelting process was also easy to produce. For the first time in history, ordinary people had the opportunity to buy and use metal, at least in small quantities. The use of iron in agriculture as a plough improved the use of the soil and enabled the expansion of arable land. This led to gradually increasing prosperity. Because iron was cheap, a relatively large part of the male population was now able to acquire weapons and armour made of metal for warfare. Farmers and herdsmen thus acquired a previously unknown fighting power, and this completely changed the previously aristocratic social structure of the age of chariots. The spread of iron processing in the Near East between 1,200 and 1,000 BC led to numerous invasions and migratory movements. For the first time, the Assyrians clearly demonstrated their inventiveness in the development of new military equipment. Between 935 and 612 B.C. the Assyrians used new military equipment and formations. They invented, for example, a complex array of equipment for the siege of fortified cities and carried a siege train. An important weapon system that was also widely used at that time is the mounted soldier. No one today can say with certainty when the custom of riding a horse originated, nor where it happened. Early depictions show Assyrian warriors on horseback. Most historians assume that steppe peoples, who profited from the "equestrian revolution" in the most spectacular way, were the pioneers of this new use of horses, their endurance and speed. The horse as a weapon system was further developed by the Persians by breeding horses that were able to carry a man in armour. Such horses were also protected with armour against arrow shots. The new weapon system spread only very slowly to Western Europe. Only in the period from Hadrian 117 to 138 A.D. did the Romans experiment with it, but the number of cataphracters was limited.

In India the first great empire, the Maurja empire, was formed around 322 B.C. King Chandragupta maintained an extensive spy organization and an excellently equipped army to secure his power. In its heyday it counted 700,000 men, 9,000 battle elephants and 10,000 war chariots. A war council of 30 officers led the operations.

In East Asia the development took a different direction. Despite Emperor Wu-ti's expedition in 101 B.C., which brought the great Persian war horses to China, these animals never gained much importance. China relied more on its powerful crossbows, whose projectiles could throw an armoured warrior at a distance of almost a hundred metres from his horse. This greatly reduced the effect of heavily armoured riders.

Middle Ages

The Middle Ages is a term coined by historians for the period between antiquity and modern times. Beginning and end are set differently: from the beginning of the migration of peoples around 375 or the fall of the Western Roman Empire in 476 or the time of Charlemagne around 800 until the discovery of America in 1492.

The time of the migration of peoples is understood to be those features of Germanic and other peoples to the west and south of Europe in the 4th to the 6th century AD, through which the Roman Empire was destroyed and the transition from antiquity to the Middle Ages was completed. How many tribes and how many people took part in the migration of peoples cannot be said with absolute certainty today. According to estimates, however, some people may have invaded western and southern Europe, settled there and mixed with the local population. The migration of peoples in Europe led to a mixing of peoples, but also to economic and cultural decline, which temporarily resulted in an exchange economy.

The Middle Ages were an epoch of many small and big wars. While the small wars were mostly tribal wars, the outcome of which contributed to the rise of certain tribes, such as the Carolingians, the great wars were transnational wars, the outcome of which caused the rise of great powers, such as Probably the greatest transnational wars were the conquest campaigns of the Arabs in the spread of Islam and the crusades of the European states to reclaim the Holy Land. In the space of just 200 years, seven crusades took place, which took up great human and material resources and thus contributed to the stagnation of the economy in Europe. Let us take a closer look at the First Crusade from 1096 to 1099 and the Crusade of Frederick I from 1189 to 1192. In the First Crusade, the Crusader army first marched through Europe in several trains and gathered off Constantinople in 1097. According to more recent estimates, the size of the army is assumed to be around 50,000 to 60,000 men, including 6,000 knights and noblemen as well as a 22,000-man infantry. The number of horses is estimated at 50,000. In 1099, after a siege of about five weeks, the conquest of Jerusalem and the subsequent establishment of feudal feudal states with independent vassals following the French model. Jerusalem was conquered again in 1187 by Sultan Saladin. This was the basis for the third crusade. In the Third Crusade, Emperor Frederick I Barbarossa placed himself at the head of the entire Western enterprise in accordance with his idea of the universal position of the emperor. Other participants were the English King Richard the Lionheart and the French King Philip II. August with their retinue. The German crusader army chose the land route along the Danube via the Balkans. They made a stop in Belgrade. From the original 25,000 men who set off from Regensburg, the strength of the army grew to 100,000 men through the addition of Austrian and Hungarian soldiers. The crusade came to an abrupt end when the emperor drowned while bathing in Cilicia in the river Saleph (today Göksu). A large part of the demoralized troops then returned home, and a part of the German crusader army, under the command of Frederick's son, Duke Frederick V, reached the Holy Land and took part in the battles for Acre. The crusades failed because the national interests of the participating nations could not be united with the universal idea.

Among the most important weapons of the time were bows and arrows, lances, swords and, from the 11th century, the crossbow. The weapons were made by craftsmen. The bows of Bognern and the arrows of arrow carvers. The bogner and arrow carvers were used, in addition to their main task, the craft activity, also for weapon service on the ring walls of the towns. When the bow was replaced by the crossbow, the bogners became crossbows. The swords were made by knife- and blade smiths, whereas cutlers specialized in the production of defensive weapons and all kinds of knives, daggers, hack knives, single-edged blades and the other double-edged blades were made by blade smiths. For the production of sword and sabre blades, a complicated technique originating from the Orient was used in the Middle Ages and later on, called dyeing or damascene. In the centres of blade forging such as Solingen, Nuremberg, Regensburg and Steyr, a separate profession developed, that of hardener, who concentrated exclusively on this work, while elsewhere the blacksmith hardened his blade himself.

After the collapse of the Roman Empire there was a stagnation in the development of armaments. Thus the chain mail remained the most important protection of the rich warriors or the nobility for a long time. The armour of the knights of the Middle Ages goes back to the armour of antiquity. They are a consistent further development of the first plate armour from the time of the Greek and Roman heyday. The plate armour was a measure against the advent of crossbows and longbows. The first effective plate armour was created in the 13th century. The protection initially covered the body and head, later the limbs. The manufacturers of plate armour were called Plattner in Germany and were organised into guilds. The Plattner bought the steel plates from large forges. The Plattner proceeded like a tailor when making the armour. He first determined the body measurements, drew the patterns on the steel plates and removed the marked parts with chisels and metal scissors. The plates were then heated, bent and brought into the desired shape by hammering. Much of the work was done on the cooled metal. Once all the parts of the armour were finished, locksmiths attached rivets, leather straps and hinges to the plates. Then the plates were polished on leather-covered wooden slides and, if necessary, decorated. As a last step, the plattner had the armour parts covered with a lining of wool or grass, which was then covered with linen, cloth or silk. A full armour (from French harnais) was an armour that protected the entire body; it weighed on average 20 to 30 kilograms. The weight of custom-made armour was evenly distributed over the body. The biggest problem with a plate armour was not its weight, but the heat generated. For example, the Duke of York is said to have died of a heart attack in the Battle of Azincourt in 1415, which resulted from the great heat in his armour. The platteners had to have a good knowledge of the human locomotor system in order to be able to produce armour that was as flexible as possible. NASA also studied the knight's armour in detail in order to get impulses for the construction of an effective space suit.

While the production facilities for armour were still small manufacturing plants at the beginning of the 12th century, from the 15th century onwards they developed into real production centres for the production of armour. At that time, the leaders were the northern Italian and southern German platteners. Italian armour was already being exported to the whole of Europe in the late 13th century. Important centres were the cities of Milan, Florence, Brescia, Genoa, Venice, Modena and Rome on Italian soil and on German soil Augsburg, Landshut, Nuremberg and Innsbruck. Smaller centres existed in Ulm, Cologne, Vienna, Magdeburg and Lübeck. Italian and German masters also established production facilities in other countries. In England, Henry VIII set up a royal smithy in Greenwich in 1515. In Scotland, King James IV had armour made in Edinburgh and his successor James V opened another forge in Holyrood in 1531. French armour was made in Paris, St. Quentin, Tours and Rouen. Eastern European armour came mostly from Krakow.

Hardly any invention influenced the warfare more than the invention of gunpowder. Based on gunpowder, new weapons could be developed whose destructive power was to eclipse any weapon system known up to that time. The spread of knowledge about the existence of gunpowder was therefore dangerous during the Middle Ages, both from a secular and a spiritual point of view. This was because the ruling forces now had to fear losing their supremacy to opponents who had enough gunpowder and weapons with which to use it. To betray the formula was equally dangerous. The Franciscan monk from Somerset, Roger Bacon, tried to prove in a writing "Epistola de Secretis Operibus Artis et Naturae" around 1249 that things that are usually attributed to more devilish magic can be imitated by human hands. He noted that a certain mixture can be made to explode if one knows the trick and gave a formula. Translated from Latin it reads: "Take saltpetre, LURI VOPO VIR CAN UTRI, and sulphur, this together gives great thunder and lightning". The letters of the anagram, here in capital letters for distinction, are translated backwards: "Take seven parts saltpetre, five parts charcoal from young neck nut wood and five parts sulphur". A formula for the production of saltpetre was also given. More precisely, this is a mixture of 41.2% saltpetre, 29.4%

charcoal and 29.4% sulphur, which is suitable for fireworks but not for firearms. At the same time in Germany, a Franciscan monk named Berthold Schwarz, who bore the civil name Konstantin Anklitzen, is said to have been engaged in chemical experiments and allegedly invented the explosive effect of a mixture of saltpetre, sulphur and mercury or of saltpetre, sulphur, lead and oil around 1259. A report from 1410 says: "This art was found by a magister. His name was Magister Berthold...who was engaged in alchemy...he mixed the ingredients in a copper crucible and sealed it tightly...and put it on the fire...the crucible shattered into small pieces...Later the Magister tried to find out if it was possible to throw a stone away with this remedy...". There is no certainty about this, but it seems to have been a widespread belief in Germany that a Franciscan monk invented gunpowder at the beginning of the 14th century. In 1853 a monument was erected to the Franciscan monk Schwarz in Freiburg. Some people today still consider the German monk Schwarz to be the European inventor of gunpowder, although for many historians it is not even certain that he ever lived.

The date for the invention of firearms can be determined with a certain degree of certainty on the basis of the first uses in Europe. We know, for example, that during the siege of Trento by the Veronese in 1278, we can be sure that during the siege of Trento by the Veronese in 1278, devices with operating specialists were used to throw iron and fire, which the Count of Tyrol lent to Verona. These new types of guns, known as "bombarda", were probably made in Austria or Italy. In France such guns were used during the siege of La Reole in 1324. From the very beginning, the firearm had a double effect when used: a technical and a psychological effect. The first extremely primitive guns were called fire vases and they were used to shoot big arrows. Between 1300 and 1350 these fire vases became more mobile and instead of arrows they were used to shoot stone balls and later on iron balls. The new types of weapons shot their load in an imprecise manner, so that the old weapons remained in the army's inventory. The guns that spit fire were feared primarily because of their thunderous bang and the smoke they produced. Field cannons were first used in the Battle of Crecy in 1346. For the sake of completeness, it must be said that these guns did not make a decisive contribution to the English victory. Of the 9,000 soldiers on the English side, 5,000 were long bowmen and these continued to decide the battles between the English and French for almost a century. Almost all such fire pits were made of iron rods welded together, surrounded and reinforced with hoops. Between 1350 and 1450 the size and low thermal efficiency of the furnaces did not allow the casting of large pieces. The cannons often burst, which was also due to the wrong charge. Over time, however, they learned to improve the quality of the powder and to measure the charges correctly. The appearance of the first hand guns can be dated to the middle of the 14th century. The existence of such firearms is mentioned around 1346 in the "English Royal Household Accounts". The oldest hand gun was found in the fortress of Tanneberg, which was burnt down, dragged and never rebuilt in 1399. When these first handguns appeared, the death bells rang for knighthood. However, their improvement was very slow and it is no exaggeration to say that centuries passed between this ringing of bells and the final death of the armoured riders as battles. The first hand guns consisted of a simple tubular stock with an ignition hole at the end. The whole thing was quite unwieldy, which is why very soon the arca bouza (arca bouza = bow with hole) appeared, which used the shaft of the crossbow and its firing methods. The bullets were mostly bullets or arrows, crossbow arrows or bolts, which were stabilized by twists that could be created by springs at the rear end or by a guide in the barrel. Already towards the end of the 15th century, the first rifles or rifles with rifle barrels appeared. Simultaneously with the development of small arms, the profession of gunsmith was born. Before that cannons and rifles were produced by bell founders, coarse blacksmiths and watchmakers. The stocks were made by wood carvers. The omnipotent guilds began to limit production and keep prices down. The only exception were the gunsmiths, who worked from private workshops for the emperors and kings. On German soil, the first gunsmith centres were formed in Nuremberg, Suhl, Augsburg and Solingen, in Holland in Amsterdam, Utrecht and Maastricht. With the advent of the wheel lock around 1515, many watchmakers switched to locksmiths and supplied the gunsmiths.

The Thirty Years' War completely destroyed German arms production, and although the Italian cities of Brescia, Milan, Florence and Gardonne produced very good weapons, very few rifles were made.

The arms industry was perfected in Milan and reached the peak of its development in the 14th century. The production of armour and weapons of all kinds in the 100 or so armouries in Milan was encouraged and favoured by the quality of the raw materials extracted from the ironworks in the Lombardy valleys and by the demand that came from all over Europe from the middle of the 13th century onwards, where entire mercenary armies had to be armed and equipped. The finely woven chain mail of these armours owed its worldwide fame to the technical skills of the master craftsmen who were in direct contact with the buyer and worked on their own account in their small workshops, relying only on the help of a limited number of workers and journeymen. While the artisanal manufacture of weapons remained in the hands of a few specialists, shipbuilding, especially in Italy, took on an industrial character from the 13th century onwards. In Genoa, the construction of the galleys required the cooperation of private and state industry. In Venice, two differently specialized branches of shipbuilding had developed. The state industry was concentrated in the famous Arsenal and served the needs of the navy, while private shipbuilding was spread over countless small shipyards, so-called "squeri".

The Modern Age

early modern times

The largest early modern weapon manufacturing complex was located in north-western Europe in the southern Netherlands and the Lower Rhine area. Extensive production structures stretched from Hainaut to the centres of Liège, Namur and Maastricht, Aachen/Stolberg and Cologne, and far into the central-west German region of Westphalia. Nuremberg had large production capacities in southern Germany. Nuremberg had specialised in the mass production of firearms at an early stage and had also become the leading German centre for gun casting and rolling brass production. In addition, Suhl in Thuringia rose to become an important European production centre for military goods in the last third of the 16th century by specialising in cheap mass-produced goods for the common soldier and through its commercial ties with Nuremberg.

Sweden under Gustav II Adolf forced the establishment of an independent arms industry by recruiting experts from Holland. With the immigration of the leading Dutch armaments expert Louis de Geer, Sweden built up extensive production structures for weapons and ammunition from 1627. Sweden established a new heavy industry at the copper mine in Falun, in Finspag and Nordköpping, as well as gunpowder and fuse production in eastern Gotland. German and Dutch specialists, mainly recruited from Liège and Aachen, were used to a large extent. In 1636 there was even an attempt to transfer the entire Aachen brass industry to Sweden.

The great military conflicts of the 16th century - one thinks of the campaigns of Charles V and Philip II, the permanent conflict with the Ottoman Empire, the Dutch War of Independence, and the large number of "local" conflicts within the European countries - created a relatively stable demand for armaments, which in turn led to an extremely favourable wartime economy and thus to excellent development conditions in the field of military production, culminating in the Thirty Years' War. This international conflict surpassed anything that had ever happened before, especially with regard to the enormous increase in the quantity of war material, the short delivery times required and the duration of the arms boom. In all European countries the demand for military equipment was so great that the increasing demand could hardly be satisfied and only a Europe-wide network of procurement structures could ensure a halfway orderly and continuously flowing supply. All available resources and relationships in the commercial and financial sectors had to be mobilised to ensure that the war machine could function smoothly. Of course, all European countries had armament structures that more or less corresponded to their real military needs. Moreover, the state stockpile system for storing war material was only just beginning to be established. In the event of war, one was forced to turn to the large European production and distribution centres for materials for military requirements. Thus, during the Thirty Years' War, the major European arms producers exported their products in all directions across all borders, regardless of political coalitions or military opposition. For example, Nuremberg repeatedly supplied war materials to the imperial enemy number one, France, via Lyon and supplied both the Catholic and Protestant sides. The political instability of the South Netherlands-Lower Rhine arms manufacturing complex was one of the main reasons for the extensive arms trade of these areas with all European countries and political parties, because it was here that the international mercantile mercantilism found the necessary trade policy space between the fronts.

The continuously increasing demand for military equipment was the prerequisite for the training, expansion possibilities and specialisation tendencies of the European armament centres, which had reached their economic peak with the Thirty Years' War.

The Technical Revolution

In the first half of the 19th century, Germany experienced a faster economic development than the leading industrial countries of that time, France and Great Britain. The share of the countries of the German Empire in the rapidly growing world economy, especially in world industrial production, doubled between 1820 and 1913. The relatively rapid development in Germany was based on the consistent exploitation of the then modern technologies of chemistry and electricity and on the transition from iron to steel. In addition, the division of labour, mechanisation and concentration continued to advance. With its rapidly growing economic power, the German Empire demanded its share of the already divided world. Germany did not want to do without the possession of colonies as a source of raw materials. The German Empire's desire for more prestige in the world required an efficient war fleet. Germany oriented its armaments economy towards the expansion of a deep-sea fleet that could at least secure the worldwide interests of the German Empire. The armament of the fleet proceeded rapidly in Germany. Beneficiaries of the expansion of the fleet were the shipyards, e.g. the Vulcan shipyard in Stettin, Blohm + Voss in Hamburg, the Germania shipyard and the Howaldtswerke in Kiel. By 1914, 38 liners and battleships, 13 large and 17 small cruisers, 150 torpedo boats and 44 submarines had been built at German shipyards.

The 20th century to the end of the Cold War

There is no comprehensive and detailed estimate of arms and war expenditures in the first half of the 20th century. Bernhard Endrucks has, unfortunately without exact source data, presented the following calculations of the extent of the armament and war expenditures including the war damages of the world in the first half of the 20th century as a whole. His research results can be seen in the figure.

Figure: World military expenditure 1900-1952 in billion gold dollars

Wars	Issues
1901-1914	52,6
Smaller wars 1901-1914	5,5
First World War	260
1918-1939	232,4
Smaller wars between 1918-1939	29,5
Second World War (including war damage)	3300
1945-1952 (including the Korean War)	777,6
1900-1952	4657,6

At the beginning of the 20th century, one of the tasks of the still young vehicle industry was to build vehicles for military purposes with which the extremely heavy artillery guns could be transported over difficult terrain. Great Britain built a large number of steam locomotives for the road, which found buyers in various European countries. Manufacturers such as Aveling & Porter or Fowler supplied the armed forces of Italy, Russia and France with their vehicles. Despite its considerable dimensions, the Aveling & Porter vehicle was relatively light - it weighed 5.5 tonnes. The boiler could develop an effective power of about 40 hp, had rear wheels with a diameter of 152 cm and front wheels with a diameter of 107 cm. According to the wishes of its inventors it should be able to transport artillery and bridge material up to a weight of six tons. With the threat of the First World War and the simultaneous development of civilian automobile transport, the interest of military commanders grew steadily.

The First World War, which broke out in the summer of 1914, also placed new demands on the automotive industry, although motorisation was not yet very far advanced. For long distances, the railway was used, in the areas near the front the horse remained the tried and tested means of transport, while the car was still unsuitable for combat operations. Even in peacetime, so-called subsidised lorries were used to ensure that means of transport were available in the event of conflict: Buyers of a war-capable truck received a subsidy, but undertake to make the vehicle available to the army in case of mobilisation. The military was thus spared the need to create and maintain a standing stock. Passenger cars are used exclusively as staff vehicles and are confiscated at the beginning of the war.

The period after the Cold War

The end of the Cold War also marked the end of the arms race between NATO and the Warsaw Pact or between the USA and the USSR. The post-Cold War period is characterised by a high-tech defensive arms race between the USA and the "rogue states" they so called. It is postulated that even an underdeveloped country with few intercontinental weapons of mass destruction can effectively threaten a highly developed country. On the one hand, attempts are now being made to prevent this threat by imposing controls on these states with regard to the production of ABC weapons. In addition, the USA is trying to develop defence systems that can destroy intercontinental weapons on approach.

Historical examples of military technology research

Military oriented research has always played an important role in the course of human history. The development of armaments often led to a range of products that could also be used for civilian purposes; one thinks of the invention of gunpowder and its further development into explosives or the inventions that made seafaring safer.

Nevertheless, it was not until the Second World War that there was a radical change towards application-oriented research, which produced a wealth of goods that had a decisive influence on technical development. The achievements of the modern high technologies, which for the most part grew out of military research and space projects, profoundly changed lifestyles in the industrialized countries in the second half of the 20th century.

Probably the most important inventions of the Second World War with a comprehensive effect on civil life are the development of rockets, jet engines, the further development of radar and the development of the atomic bomb, which gave nuclear research an immense boost and thus made the civil use of nuclear energy acceptable. Let us take a closer look at some of these developments.

The Development of the Rocket Weapon by the German Reich

The Germans had surprised the world in 1918, the last year of the First World War, when they deployed their Paris gun with a range of more than 120 km. The French were so impressed by the gun that at their instigation it was included in the Treaty of Versailles that the Germans were forbidden to build such guns in future. This provision of the Versailles Peace Treaty was the basis for the Germans to start developing rockets in the 1920s to circumvent the provisions of the peace treaty. The young scientists who developed rockets were soon joined by Wernher von Braun, who was to become the leading development manager of the German rocket program. All rockets were initially given the designation A for aggregate. The A1 rocket did not go beyond the project stage, only the type A2 was developed from 1933 and also built for experimental purposes. After the missile test centre was moved from Kummersdorf near Berlin to Peenemünde on the Baltic island of Usedom in 1937, the Type A3 was developed and tested. The experiences with the A3 were the basis for the A3, which was to become the standard long-range rocket of the German Army under the designation V2. In 1942, after the elimination of initial problems, launch attempts were made; only the fourth attempt was a complete success. From 1943 the rocket went into series production. On 8 September 1944 the first V2 was launched on England. A total of 1115 V2s were shot down on England. Their result achieved exactly the effect expected by the client. The 1 ton of explosives caused considerable damage in a wide radius. Besides England, Antwerp and Liège were also Antwerp and Liège after the landing of the Allies and the occupation of large parts of Western Europe. In addition to the A4, rockets were also planned which would have been able to reach New York. However, the project had not progressed beyond the drawing board stage by the end of the war. The A9/A10 was planned as a two-stage rocket. The larger stage was used to launch the composite rocket and send it into the stratosphere. There the second stage was ignited, it then ejected the launch rocket. On 2 May 1945 Wernher von Braun surrendered to the US Army and was brought to the USA together with other scientists. About 100 booty specimens of the V2 were dismantled by the USA in Nordhausen, the place of manufacture of the V2, and sent to the USA. These rockets formed the basis for the US rocket and space program. One of these specimens can still be seen today in the National Air and Space Museum in Washington. The further relocation of renowned German rocket scientists in the summer of 1945 took place under Operation Overcast. The USA extensively tested the German rockets first in New Mexico, then in Huntsville, Alabama, so that very soon around 1950 routine rocket launches could take place from Cape Canaveral in Florida. Just like the USA, the Soviets had brought the remains of German rocket technology and available scientists to the Soviet Union. The Soviet R-1 rocket was the replica of the V2. It was first launched in 1947 from the test site Kapustin Jar. It should only be mentioned here in passing that German rocket technology was very advanced and rockets for air defence had already been developed, such as the Rhine Daughter, Butterfly, Waterfall and Gentian.

The development of the jet engine

When on 15 May 1941 the British Gloster took off from the ground for the first time with a Whittle W1 jet engine, the Allies were convinced that they had launched the first jet aircraft of the war. Only after the war did it become known that the Germans had already been two years ahead of them. On the German side, the development of the jet engine began with Ernst Heinkel. Ernst Heinkel won Hans Joachim Pabst von Ohain (1911-1998), who developed a jet engine in 1937. Ernst Heinkel had a small aircraft built, the He 178, which was equipped with the jet engine. In June 1939, the He 178 took off successfully and landed safely again. This heralded the age of aviation with aircraft equipped with jet engines. After the initial success, the jet engine was quickly put into series production, but it was to last until 1944, when the Messerschmitt Me 262 went into series production. By the end of the war, 1400 machines of this type had been built. In 1947, Hans von Ohain, like many other German engineers, was brought to the USA by the Americans with inventions relevant to military technology as part of Operation Overcast. At first he worked for the Air Force and supported them in the development of their own jet aircraft. In 1956 Ohain became director of the Air Force Aeronautical Research Laboratory and in 1975 he was promoted there to chief developer of the Aero Propulsion Laboratory. Together with Frank Whittle, Ohain received the "Charles Stark Draper Prize" in 1991 for his pioneering developments in the field of jet engines. After the war, the development of the jet engine not only influenced military aviation but also revolutionized passenger aviation.

Helicopter

The idea of vertical take-off goes back to a Chinese toy that was developed around 2000 years ago. But it was not until the 20th century that people were technically able to build fully-grown helicopters. In the beginning, these rotary-wing aircraft were still bouncing and hovering flying machines, but from the mid-1930s onwards, German engineers made the breakthrough. The German Reich was also the first nation to produce a helicopter in series. This was the Flettner Fl 282 Kolibri. During the war at least 20 of these were delivered to the armed forces until 1943; most of them were used for reconnaissance. Over the big pond, in the USA a young Russian exile with the well known name Igor Sikorsky Helicopter tried to develop. After several prototypes Sikorsky developed the R-6A in 1943, which flew for the first time in October 1943. With this helicopter, which already showed characteristics of modern helicopters - the streamlined tadpole fuselage and a Plexiglas bubble as cockpit - the rapid development began after the war, although from a purely technical point of view the German helicopters were more advanced than the helicopters of Sikorsky. This was recognized by the combat value of the helicopter in use in Korea, Malaya, Algeria and finally in Vietnam. Both in the West, Sikorsky, Bell, Hughes and Kaman in the USA, Bristol Aeroplane Company in Great Britain and SNCASO in France, as well as in the East by Mil and Yakovlev, the development of military and civil helicopters was pushed forward.

Innovations in shipbuilding

The Navy, as one of the branches of the armed forces, has always been a testing ground for technicians and engineers, but it was only during the arms race before the First World War that naval engineers came up with the decisive innovations that revolutionized the use of ships. As an example, the development of propulsion systems will be discussed in more detail. Around the beginning of the 20th century, the steam engine had been pushed to the limits of its performance. Neither speed could be increased nor reliability improved. In order to increase the speed of the increasingly heavy combat ships, new propulsion systems were sought. The English engineer Sir Charles A. Parsons took a decisive step forward by developing the overpressure steam turbine. Without wanting to go into the technical details, this development revolutionized the possible applications of combat ships. The H.M.S Dreadnought, which during its test voyage from Portsmouth to the Mediterranean Sea and on to Trinidad and back to England at a cruising speed of 17.5 knots without any engine damage, achieved the impressive proof of the new turbine's power - a performance that no ship with piston steam engines had ever achieved.

Other important innovations in warshipbuilding that have influenced civil shipbuilding are the switch from coal-fired to oil-fired ships and finally to nuclear propulsion, navigation systems, composite materials and sonar systems.

The development of the radar

Aerial warfare played a decisive role in the Second World War. Although the expected psychological effects of bomb warfare on the enemy civilian population did not show the results that its supporters hoped for, the war of the Allied bombers against military and civilian targets in the German Reich helped to ensure that the German Reich was not able to produce armaments without borders. The success of the bombers was only possible, however, because of the great success of research and development in the field of radio communication and radar technology. Germany and England were already working on electronic navigation aids for the Luftwaffe during the 1930s. The English were working on a radar defence system. Thanks to the preliminary work of physicist R.A. Watson-Watt, the British developed a radar which was installed on the east and south coast of England at the beginning of the war. They can detect aircraft as far away as 160 km. During the air battle, these radars were an important advantage for the British air defence. The Germans developed two types, the precision radar "Würzburg" and the early warning device "Freya". In addition to the development of the radar, radio navigation procedures were also further developed. Inventions today are indispensable in military and civil road, sea and air traffic.

The age of nuclear energy begins

In 1939, Albert Einstein recommends to US President F.D. Roosevelt the development of an atomic bomb. In the course of the war a race for the super weapon begins. On December 2, 1942, some nuclear physicists who had emigrated from Europe set the first chain reaction in motion in the USA. The nuclear reactor stands under the stands of the football stadium of the University of Chicago. The preparations were carried out under the strictest secrecy. The success of Chicago spurred on those responsible in the USA. A gigantic project began in the New Mexico desert. On July 16, 1945, the world's first atomic bomb was detonated. After that, only a few weeks were to pass until the new weapon would be used for the first time in war. The dropping of atomic bombs on Hiroshima and Nagasaki accelerated the end of the war in the Far East and secured the USA a brief monopoly (until October 1949) of this new weapon. It need not be emphasized that World War II accelerated the development of nuclear technology.

The historical examples only provide an insight into the many different ways in which military research can be used in the civilian sector. Looking at modern research and development in the field of military equipment at the end of the 20th century and the beginning of the 21st century, it is not difficult to see that information technology, the materials industry and also robot technology receive decisive impulses from military research.

The computer and the Internet

The first functioning computers were used by the military. From the beginning, the military use of computers has been one of the main driving forces behind computer development. During the war (1941) Konrad Zuse built the first functional program-controlled binary calculating machine. The next digital computers were the Atanasoff-Berry computer built in the USA (commissioned in 1941) and the British Colossus (1941).

Computers are now used to plan and execute military operations, analyzing reconnaissance data in the same way as a tank's weapons computer. But Shannon's prognosis proved to be too optimistic: strategic and also tactical decisions in the military are still largely made by humans.

The history of the Internet can be divided into three phases. In the early phase from the mid-1960s onwards, the foundations were laid, the technology was demonstrated and developed to the point of usability. The success story of the Internet begins with the Arpanet. The ARPANET (*Advanced Research Projects Agency Network*) was a computer network and was originally developed on behalf of the US Air Force from 1968 by a small group of researchers under the direction of the Massachusetts Institute of Technology and the US Department of Defense. It is the precursor of today's Internet. In the late 1970s, at the same time as the change from military to academic research funding, the growth and international spread of the Internet began. This period saw what is commonly associated with the wild phase of the original Internet: an economy of exchange for software and information, grassroots-based self-organization, evolving communities, and the hacker spirit that knows how to circumvent any restrictions on access and the free flow of information. The commercial phase of the Internet began in 1990 with the shutdown of Arpanet.

The artificial intelligence

On 9 March 1949, at the annual meeting of the Institute of Radio Engineers in New York, Claude E. Shannon presented a paper on programming a computer to play chess. Although this was probably of no practical significance, the question was of theoretical interest and could help solve similar problems of greater importance, wrote Shannon, who became known as the "Father of the Information Age". He also spoke of "mechanized thinking", which could, for example, help to develop machines "to make strategic decisions in simplified military operations".

Artificial intelligence (AI), as mechanized thinking has been called since 1955, is now considered the fourth industrial revolution. It changes many areas of the economy and life once again fundamentally. It will also not leave the military power relations and the geopolitical constellation untouched.

While the German Reich developed the jet comb plane Me 262 ready for use,

the USA succeeded in developing the atomic bomb ready for use.

National defence industry as a research and technology carrier

Among economists, there has been a heated debate for years on whether spending on research in the military goods producing sector is slowing down the growth of an economy. The argument is based on the opportunity costs that arise from the use of expenditure in this area of research, because this deprives funding for more efficient research in the consumer goods producing sector and thus the supply of consumer goods can no longer keep pace with demand, which ultimately has a negative impact on economic growth. This academic discussion, although extremely interesting, will not be dealt with in detail in the following, because wars seem to be part of human nature like laws of nature and are fought with weapons that are constructed in an increasingly sophisticated way and developed with ever greater research effort. From the author's point of view, this leads to the conclusion that as long as there are wars, weapons will be used and also produced.

If we look back at the images of the last Iraq war in 2003, it seems that the fact that qualitative nuances of technical progress decide on victory and defeat is of decisive importance. Furthermore, the pictures only too clearly show us the obvious technological advantage of US armament over Europe. Technical progress has long been regarded as the main engine of economic development, and especially for the growth of an economy, at least since the studies of Robert M. Solow, the 1987 Nobel Prize winner for economics. It is not surprising, therefore, that countries that pay particular attention to nation-state research can also look back on impressive economic development and achieve annual economic growth from a very high starting level. In connection with the importance of research and development, the method of research and development should also be seen in a different light from country to country.

If we look, even if only superficially, at the global importance of research and development (R&D), the OECD countries spent some US$ 645 billion on research and development in 2001, 44% of which was spent in the United States of America, 28% in the European Union and 17% in Japan. Over the period 1995-2001, the OECD average annual growth rate of R&D expenditure was 4.7%. Again, the United States led the way with around 5.4%, followed by the EU with 3.7% and Japan with 2.8%. Sweden, Finland, Japan and Iceland were the only OECD countries where R&D as a percentage of GDP exceeded 3% in 2001. It is worth noting that the increase in investment in research is mainly due to the increase in R&D expenditure in the private sector. Government spending on R&D has been declining in recent years, partly due to the decline in defence R&D and the privatisation of certain public institutions. Nevertheless, it should not be overlooked that Europe has excellent research facilities and is also capable of producing excellent high-tech products. However, Europe still lacks the bundling effects needed to ensure more efficient product manufacture from networked research facilities as a whole.

In connection with the alleged technological superiority of the USA in military equipment, it is worthwhile to compare research and development expenditure on armaments, provided that the openly accessible figures allow a serious comparison at all. If one shares the view of scientists that investment in research is a major driver of technological development, it is not surprising that the US has assumed technological leadership in defence equipment, as it spent some US$ 70 billion on R&D for defence and internal security in 2004, while European spending on defence research was only one-fifth of US spending. And the gap will widen in the coming years as the US steadily increases its defence research budget. With the financial resources available, the US is able to provide its top 10 defence companies with considerable sums of money for R&D. With the research subsidies granted by the government and the company's own investments in R&D, the US armaments companies can thus maintain their technological lead over other countries. This, in turn, secures sales markets and revenues from defence equipment business. The superiority of US armaments is reflected in a ranking of the 100 commercially most successful defense companies:

Table: Defence business in 2019 of the top 100 placed defence companies

Country	Number of Company	Turnover in billion US$	Comments: Top-selling companies/turnover in US$ billion
USA	40	260	Lockheed Martin 50, Boeing 34 Northrop Grumman 25, General Dynamics 24
PRC	8	100	Aviation 25, China North 15, Aerospace 12, China South 12, Shipbuilding 10
Great Britain	10	40	BAE Systems 22; Rolls-Royce 4.6
France	5	23	Thales 9.5; Naval 4.2
Netherlands	1 (Airbus)	13	-
Italy	2	12	Leonardo 9.8
Turkey	4	5	Aerospace 1
FRG	3	8	Rheinmetall 3.4
Israel	3	10	Israel Aircraft 2.6; Rafael 2.5; Elbit 3.3
Russia	2	15	Almaz 9.6
Other	22	24	
		510	

Source: Defence News 2020

As can be seen from the table above, 40 US companies account for around 55% of the total turnover of the top 100 defence companies, while European and Asian companies each account for just over 20% of turnover. The rest of the world is insignificant in terms of defence equipment production.

IV. Country reports

4.1 Smaller countries without significant defence industries

Albania

The ancestors of today's Albania settled around the 8th century before Christ. In Greek and Roman times, the country played an important strategic role at the entrance to the Adriatic Sea. With the migration of peoples, the Goths advanced as far as Albania. Since the 6th century the Slavs have invaded the country. In the Middle Ages Albania was fought over by Serbs, Bulgarians and Normans. Since 1385 the Turks invaded and conquered the country. Until the beginning of the 20th century, the population rebelled against the Turkish occupiers on several occasions. Only in 1912 they were able to shake off the yoke of occupation. During the First World War, Albania was briefly occupied by the Central Powers. After the First World War, Albania became increasingly dependent on Italy; it was finally annexed by Italy from 1939 to 1943. After the Second World War, Albania, under the former partisan leader Enver Hodscha, was under Yugoslav influence until 1948, became a People's Republic in 1946 and sought Soviet support from 1948. Since 1956, however, a rapprochement with the People's Republic of China took place. Albania finally withdrew from the Warsaw Pact in 1968. Albania has been a NATO member since 2009 and has limited capacity to produce military goods such as explosives and ammunition, small arms and machine guns and naval ordnance.

Andorra

Andorra is a political curiosity. For 1000 years the small state in the Pyrenees has been able to maintain its independence between France and Spain. Andorra has no armaments industry.

Bosnia and Herzegovina

Bosnia had been under the alternating sovereignty of the Serbs, Croats, Byzantines since the 10th century and the Hungarians since the middle of the 13th century. In the middle of the 14th century, under the dynasty of the Kotromanici, Bosnia broke away from Hungarian dependence; it gained the Banat Rama, which was renamed Herzegovina (=Dukedom). In 1377 Stephan Tvrtko was crowned king of the Serbs, Bosnians and Dalmatians. In 1400 Hungary regained Dalmatia and finally the Turks conquered Bosnia around 1463. 1908 Austria-Hungary annexed Bosnia Herzegovina.

In the territory of today's Bosnia-Herzegovina, mainly in the Serbian inhabited part, about 42% of all armament enterprises in the former Yugoslavia were located with a total number of 38,000 employees. Important centres of the armament industry were Banja Luka, Novi Travnik and Mostar, but small enterprises were also located in Gorazde, Konjic, Sarajevo and Bratunac. After the division of the state as a whole, there was a conversion of the enterprises, with about 70% of the arms production being reorganized into civil-goods producing enterprises. At present, about 7,000 workers in 9 armament factories produce Unis Ginex in Gorazde, Binas in Bugojno, BNT in Novi Travnik, TRZ in Hadzici, Igman in Koncjic, Zrak in Sarajevo, Unis Pretis in Vogosca and Vitezit in Vitez. In the Autonomous Republic of Srpska 6 large companies produce Famos in Cajevac, TRZ in Bratunac, Pretis, Orao in Bijeljina and Kosmos.

After the reorganization of the state, efforts are being made to get an armaments industry going again. Ammunition factories (Igman and Pretis) and a production facility for grenade launchers (BNT-TMiH) are already back in operation, as are repair companies for land (TRZ) and aircraft (ORAO) and electronics (ZRAK).

Estonia

The history of Estonia and its population goes back a very long time; Estonians have been living in their present settlement areas for about 5000 years. The Baltic States, of which Estonia is a part, have been a plaything of the great powers since the 12th century. German, Danish, Swedish, Russian and Polish domination influenced Estonia until 1920, followed by a period of national independence which lasted until 1940. After that Estonia was part of the Soviet Union until 1991. In 1991 the country became a member of the United Nations, in 1993 of the Council of Europe. Since 1999 the country has been a member of the World Trade Organization (WTO). Estonia has been a member of NATO since 2004 and has a low capacity for arms production. In the field of electronics there are eight small and medium-sized companies, the Tallinn shipyard is a leader in the field of naval armament, and in land systems Baltic Defence and Technology is a company that repairs land systems and Milrem Robotics develops unmanned systems.

Ireland

Around the 3rd century before Christ, immigrating Celts displaced the original population. The Celtic upper class divided their clans into five kingdoms; the king was crowned in the common capital. Uniform law and common religion united all Irish. In contrast to England, Ireland never belonged to the Roman Empire. In the first centuries after Christianity, the Roman Church sent missionaries to the island, including the national saint Patrick, who is still venerated by the Irish today. Patrick came to the island in the 5th century and founded the Irish National Church. The so-called golden age came to an abrupt end with the Norman invasions in the 9th and 10th century. Although the Normans were defeated, there was an internal unrest afterwards. These disturbances caused Irish nobles to seek refuge in England. The English King Henry II then landed in Ireland and subjugated it. Thus began the English colonization of Ireland. The Irish often tried to shake off English sovereignty, but the English declared Ireland's independence in 1916. After renewed years of fighting, England surrendered in 1921; Ireland became a free state after the cession of 6 provinces in the north (Northern Ireland). During the Second World War, Ireland remained neutral and withdrew from the Commonwealth in 1949. Ireland has only a small capacity in the defence industry. In particular, four companies supply electronics and one company each specialises in the development of supply parts for land systems as well as air systems.

Iceland

Iceland can pride itself on having the oldest parliament in the world: in 930 the Althing, the legislative assembly of the Icelanders, was established in Thingvellir. At this point, on 17.6.1944, after foreign rule, the islanders proclaimed an independent republic. The foreign rule goes back to the year 1262, when Iceland submitted itself to the Norwegian King. The Norwegian king guaranteed the existing national laws and a regular supply of goods. Around 1380, the country was annexed to Denmark together with Norway. In 1662 Iceland recognised Danish absolutism and was administered by Danish officials. An independence movement developed against Denmark, which eventually achieved internal autonomy. In 1918 Iceland became an independent kingdom, in personal union with Denmark. After the occupation of Denmark by German troops, the Althing transferred the king's powers to the Icelandic government. In 1940/41 Iceland was occupied by British troops and then by the USA until 1946, who kept their base in Keflavik even after that. In 1944 the political connection with Denmark was severed. Iceland joined NATO in 1949, and in 1951 it signed an agreement with the USA, entrusting them with the defence of the strategically important country. Since 1972 there has been a free trade agreement with the EC/EU. Iceland does not have an armaments economy.

Croatia

Today's Republic of Croatia comprises a territory that has been inhabited by the Croatian population since the 7th century, but has historically belonged to different political communities. From the 9th century there was a principality of Croatia, from the 10th century a kingdom of Croatia. After the rise of Hungary and Venice around 1000, Croatia came between the fronts. From 1102 to 1301, Croatia was united in personal union with Hungary, and a Croatian ban was deputy to the King of Hungary, Croatia and Dalmatia from the beginning of the 12th century. After the Habsburgs had acquired Hungary in 1527, Croatia was de jure part of the Habsburg lands until the disintegration of Austria-Hungary in 1918, but was at times ruled by the Turks. Croatia is a member of NATO since 2009.

Defence Industry

Before the War of Independence in the early 1990s, around 7% of Yugoslavia's former arms industry was located in the territory of today's Croatia. After independence from the state, the armed forces and the arms industry were reduced in size. At present, mainly companies that produce mainly civilian goods also produce military goods. According to official sources, 1,500 employees in about 25 companies produce all kinds of armaments. One of the most important companies is Đuro Đaković Specijalna vozila in Slavonski Brod, which is a subsidiary of the Đuro Đaković holding and produces armoured vehicles. Its products include the Degmann main battle tank, an improved version of the M-84A and M-84A main battle tanks, the RM-KA-02 demining machine, the LMLRS M93A3 rocket launcher, and a modernization package for the M-72 main battle tank. Another product of this company is a recovery tank mounted on the chassis of the M-84A main battle tank. According to international sources HS Product Company produces small arms. Furthermore, there are about 40 small and medium-sized companies, which primarily produce dual-use goods, which can also be used for military purposes.

Latvia

The present territory of Latvia and its population were exposed to foreign domination for centuries. From the end of the 12th century until 1991, with a short period of independence between 1920 and 1940, the country was under the sovereignty of German, Swedish, Polish and Russian. In 1991 Latvia became an independent state again. Due to the negative experiences with a policy of neutrality and the fear of Russia, which was also historically conditioned, Latvia made full integration into the political, economic and military structures of the West the goal of its foreign policy. Latvia's relations with Russia are tense because of the question of the treatment of Russian-speaking minorities. Latvia has been a member of NATO since 2004. It has about a dozen companies producing military equipment, including five companies producing electronic warfare equipment, such as UAV Factory. Of some importance is the shipyard in Riga, which has capacities for repairing combat ships. There are also two munitions factories.

Liechtenstein

The foundation of Liechtenstein dates back to 1719, when Emperor Karl VI raised the County of Vaduz and the Lordship of Schellenberg to the status of an imperial principality. Since the dissolution of the German Confederation in 1866, Liechtenstein has been an independent neutral state. After the First World War, Liechtenstein corrected its foreign policy; its strong ties to Austria were abandoned in favor of strong ties to Switzerland. Although conscription is compulsory, there have been no soldiers for over 100 years. Liechtenstein has no arms industry.

Lithuania

Lithuania has a long tradition as a nation state. The Grand Duchy of Lithuania, the last pagan country in Europe, stretched in its largest extension from the Baltic Sea to the Black Sea. Only in the middle of the 16th century did the country merge with Poland. After the division of Poland Lithuania came under the rule of the Tsarist Empire. A short phase of independence between 1917 and 1940 was followed by Lithuania's annexation to the Soviet Union. In 1991 Lithuania declared its independence and received international recognition. Since then, the main goal of Lithuanian foreign policy has been to integrate Lithuania into European and transatlantic structures. Both economic and security policy considerations have played a central role in this process. Lithuania has been a member of NATO since 2004 and has limited capacities for the production of military equipment. There are four companies that repair aircraft, the ammunition factory Giraites Ginkluotes Gamykla, as well as several electronics companies and textile factories.

Luxembourg

"Whoever has not seen Luxembourg," Goethe wrote in 1792, "will have no idea of this war building, built one on top of the other. And he continues in his description of the fortified city of Luxembourg: "From this, a chain of conspicuous bastions, redoubts, half moons, and such pincers and scribblers was created, as only the art of defence was able to do in the strangest case. A visitor to present-day Luxembourg rightly asks himself against whom this mighty fortification mountain range was built. The history of this small country begins in 963, when Count Siegfried built the Lucilinburbuc fortress. The dominion often changed hands. Among the owners are the Burgundians and Habsburgs. After the Congress of Vienna, Luxembourg became a Grand Duchy and a member of the German Confederation. The King of the Netherlands was subsequently also Grand Duke of Luxembourg. From 1890

Adolf of Nassau-Oranien was raised to Grand Duke of Luxembourg. This dynasty still exists today. 1867 is considered the year of independence, the year in which France and Prussia gave up their claims to Luxembourg in the Treaty of London and withdrew the Prussian soldiers from the casemates of the city of Luxembourg. During the First and Second World War, German troops moved into Luxembourg. Luxembourg has no capacity for the production of military equipment.

Malta

The name of the island is perhaps derived from a Phoenician word for harbour. The archipelago was already inhabited more than 6000 years ago. In this book only the most important historical events shall be presented. The name of the island probably goes back to the Phoenician name for harbour. The archipelago was already inhabited 6000 years ago. Because of its central position in the Mediterranean and its natural harbour, it has played an important role for the Mediterranean people throughout history. In the course of the millennia, Malta changed hands several times. Significant for history was the year 1530 when Emperor Charles V awarded Malta to the Order of St. John. Malta thus became a Christian outpost in the Mediterranean. The Knights, now called the Maltese, developed the island into a fortress. Napoleon destroyed the order's state. Malta then went under the protection of Great Britain. In the Peace of Paris in 1814, Great Britain was granted Malta and became a British colony. Only in 1964 Malta became independent. Since 1974 Malta is a republic. Malta has no capacities for the production of military equipment.

Macedonia

The former Yugoslav Republic of Macedonia declared independence in 1991. The territory of the state is an old settlement area, into which Slavic tribes immigrated from the 6th and 7th centuries. Today the Slavs form the majority of the population, but about 24% of the population is Albanian.

Weapons production in Macedonia is currently limited to small arms, missile launchers and ammunition.

Moldova

Moldova has been an independent republic since 1991. In addition to the majority of the population (65% Moldovans), 13% Russians and 14% Ukrainians live on the national territory. Since independence, the contrasts between the individual population groups have become more pronounced. The territory of today's Moldova is an old settlement area, which has become a transit country for various tribes since the migration of the peoples. Moldova has no capacity for arms production.

Montenegro

Montenegro is a country in the Balkans, on the Adriatic Sea. It borders Croatia and Bosnia and Herzegovina to the southeast, Serbia to the northeast, Kosovo to the east and Albania to the southeast. The Adriatic Sea is in the west of Montenegro. 1356 Montenegro becomes an independent principality Montenegro successfully resists the onslaught of the Ottoman Empire and retains a high degree of autonomy. 1799 the Ottoman Empire recognizes the independence of Montenegro 1910 Montenegro becomes a kingdom. After the I. After World War I the king is deposed and Montenegro becomes part of the new state of Yugoslavia. In 1941 Wehrmacht troops occupy Montenegro. 1945 the state of Yugoslavia is re-established. Montenegro becomes part of the socialist republic of Yugoslavia under state founder Tito. In 1992 Serbia and Montenegro form the Federal Republic of Yugoslavia. On 21 May 2006 the citizens of Montenegro vote for independence from

Serbia. Montenegro has little capacity for the production of military equipment. Tara Aerospace and Poliex should be mentioned.

Monaco

The state of Monaco owes its origin to the Phoenicians who, from Marseille, arranged the Potus Herculis Monoeci, that was conquered by the Romans in the year 154 before Christ. In the Middle Ages, the fortress and port played a role in the conflicts between Guelphs and Ghibellines. The aristocratic Grimaldi family from Genoa was among the followers of the Guelphs. Francesco Grimaldi succeeded in 1297 in conquering Monaco, which was defended by the Ghibellines. In connection with this conquest, a legend was created which is reflected in the Grimaldi family crest: the crest shows two monks armed with swords. It is said that Francesco Grimaldi, disguised as a monk, won the fortress in a coup d'état with a band of faithful. But the consolidation of the dominion lasted until the 15th century. Only then the Grimaldi could call themselves the Lords of Monaco. In 1659 the Grimaldi were raised to the rank of princes. Monaco is closely related to France and there is an economic community between Monaco and France. Monaco has no capacities for the production of armaments.

San Marino

The state of San Marino is probably the oldest republic in the world. The state was formed around 885 and since then the state has been able to defend itself successfully against all attempts at conquest. Monaco has no capacities for the production of armaments.

Slovenia

The history of Slovenia has been closely linked with that of the duchies of Carinthia, Carniola and Styria since about 900. The Republic of Slovenia has been an independent state since 1991. This was the first time that the

Slovenes achieved their own nation state. Slovenia belongs to the UNO and the Council of Europe and is a full member of NATO and the EU.

Defence Industry

The core of the Slovenian defence industry is mainly located in the regions around Ljubljana, Marburg, Ravne and Koper and is active in seven production sectors.

In the region around Ravne in Koroskem Armas produces armoured vehicles (Valuk 6x6) and artillery systems. The company also specializes in the modernization of M-84 and T-72 main battle tanks, and a new company is Valhalla, which develops weapon stands for combat vehicles.

Special attention is paid to the production of men's equipment and simulators. Slovenia does not have an efficient aviation industry. Nevertheless, an Unmanned Aerial Vehicle is produced at C-Astral.

Vatican City State

The Vatican City State emerged from the Papal States founded in the early Middle Ages, of which it is the successor state by the Lateran Treaties of 1929. The state is comparable to an absolute monarchy. The Pope has the supreme legislative, executive and judicial power. The Vatican State does not produce armaments.

Cyprus

Cyprus has been much contested due to its maritime strategic position. In the 14th century BC it was a centre of Mycenaean culture, around 1000 BC Phoenician, then Assyrian, Egyptian, Persian. In 333 B.C. it came to the empire of Alexander the Great, then belonged to the Ptolemaic empire until 58 B.C. After that it fell to the Roman Empire, later to the Byzantines, temporarily to the Arabs. The Crusader rule was followed by the rule of the Venetians. In 1571 Cyprus was occupied by the Turks and finally in 1878 by Great Britain. In 1925 it became a British crown colony. On 16.8. 1960 the declaration of independence followed. In 1974, the Turkish occupation of the northern part of the island caused it to be divided into a Greek southern part and a Turkish northern part. Cyprus does not produce armaments.

The smaller states in Europe are often dependent on aid deliveries: for example, Estonia received an older ship from Denmark.

Croatia inherited important arms industry enterprises from the former Yugoslavia, such as Đuro Đaković

4.2. Belgium

History

Belgium has only been an independent state since 1830. Before that, the territory of today's Belgium was settled by Celts and Germanic tribes and later as province Belgae part of the Roman Empire. After the migration of the peoples, the individual parts of the country experienced different fates and came under the rule of different noble houses. First the Dukes of Burgundy and later the Habsburgs ruled over the entire territory of present-day Belgium. It was not until 1830 that Belgium gained independence through an uprising and the rejection of the Dutch attempts at integration and has been ruled by kings ever since. Belgium got into the chaos of war during both the First and Second World Wars and was occupied by Germany. After the Second World War, Belgium joined NATO and is also a member of the European Union. The Belgian capital, Brussels, is home to NATO headquarters and important EU institutions.

The defence industry

As already mentioned at the beginning, the Fabrique Nationale (FN) Herstal with its headquarters in Liege is the most important company in the defence industry in Belgium. FN was founded in 1889 to fulfil an order for the production of 150,000 Mauser infantry rifles for the Belgian armed forces. Since then it has been producing weapons and weapon systems for armed forces, hunters, sport shooters and other private customers under the brand names Browning and Winchester for both the military and the civilian market. The company is able to offer a wide range of handguns, machine guns and automatic cannons including the necessary ammunition. The latest offering is the Arrows weapon system, a high-convenience weapon system developed jointly with Oerlikon. Another   major supplier of weapon systems and ammunition is Mecar, a subsidiary of the Allied Defence Group headquartered in Washington DC, which manufactures hand grenades and rifles as well as ammunition for battle tanks, grenade launchers and artillery. Its customer base includes not only the Belgian Army, but also other NATO countries, as well as non-NATO member states. A smaller supplier of combat vehicle parts is the Varec company, which manufactures parts for US combat vehicles and the Swedish CV-90 track pads, among others.

In times of surplus defence material, companies that purchase, modernise and resell old defence material can also make large profits. The SABIEX company belongs to this group of companies; so far they have bought M-109 and AMX family armoured vehicles and sold the modernised weapons all over the world. It also markets the knowledge gained from modernization and offers simulators.

The CMI company can look back on a long tradition when it took over the Fonderie de Canon de Liege. The company is regarded as one of the most important producers of weapon systems for light armoured combat vehicles. The product range extends from 90 mm to 105 mm. The company also offers system solutions for modernizing existing systems, such as retrofitting the British Scorpion or retrofitting the French AMX-13 fighter tanks with the Cockerill Mk3 or Mk8 90mm CMI system. The sister company EMI also carries out final assembly of Pandur wheeled tanks for the Belgian armed forces.

Belgium has a numerically strong air force. SABCA (Sociétés Anonyme Belge de Constructions Aéronautiques) plays an important role in the modernisation of the air fleet and has so far offered its expertise to twelve air forces. Another company in the aviation industry is Sonaca.

In the field of radar technology, the company Belgian Advanced Technology Systems (BATS) offers surveillance and medium-range radar systems and complete solutions for border surveillance.

There are also ammunition factories and textile companies that produce special military clothing.

4.3 Bulgaria

History

The first Bulgarian Empire was founded in 680 AD. From 1241 the empire disintegrated and the Ottomans conquered the area. The Bulgarians' struggle for freedom brought forth a new Kingdom of Bulgaria in 1908, which fought on the side of the Central Powers in World War I and on the side of the German Empire in World War II. In 1944 Bulgaria was conquered by the Soviet Union. In 1946 the People's Republic of Bulgaria was proclaimed. In the wake of the change of system in Eastern Europe, the resignation of the long-time head of party and state in 1989 cleared the way for reforms. Since then Bulgaria has been on the road to democratization. On the path of democratization, two orientations determine foreign policy. Bulgaria has been a member of the EU since 2007 and of NATO since 2004.

Defence Industry

At the peak of its economic success in the mid-1980s, the Bulgarian arms industry, which employed some 115,000 people, achieved a turnover of up to 1.3 billion US dollars. After the fall of the Iron Curtain and the abolition of WAPA, the arms industry now has only 25,000 employees and generates sales of around 200 million US dollars. Exports went mainly to India and Latin America.

Since 2011, the former state arsenal has been privatised and, with around 4,000 employees, produces light weapons and artillery systems, anti-aircraft guns, grenade launchers and rocket launchers including the necessary ammunition. Other production facilities for light weapons and ammunition will be NITI in Kazanlak, VMZ in Sopot, Arcus in Lyaskovets. Terem and Avionams are responsible for the repair of aviation equipment. Naval ordnance is repaired at MTG Dolphin. There are also several electronics companies.

4.4 Denmark

History

The history of Denmark is closely linked to the history of the other Northern European countries. In addition to the Scandinavian aspect, two other aspects have influenced the history of Denmark: relations with the mainland and the Viking trains. After the 10th century, when the Danish small kingdoms were united, Denmark began an expansive policy that led to countless wars. Only with the Peace of Roskilde in 1658 did Denmark give up its expansive policy, in which the state gave up its possessions in Sweden. Denmark was then internally reorganised and devoted itself more to the administration of its colonial empire, which stretched from Iceland and the North Cape to the Elbe, than to taking part in the wars in continental Europe. Nevertheless, the country was drawn into the turmoil of the Napoleonic War and had to accept the loss of Norway in the Peace of Kiel. In the 19th century, the Schleswig-Holstein question strained the relationship with Germany, which finally led to the war of 1864, in which Denmark lost to Prussia and Austria and had to cede Holstein, Lauenburg and Schleswig. In foreign policy Denmark pursued a course of neutrality. During the First World War Denmark remained neutral, but thanks to a referendum after the war it was awarded North Schleswig. During the Second World War Denmark was defeated by

German troops and quickly occupied. After the Second World War, Denmark joined NATO and is also a member of the European Union.

Defence Industry

If one looks at the equipment of the Danish armed forces over the last 60 years, it is easy to see that heavy weapon systems come from foreign weapon smiths. Denmark has a long tradition of importing larger weapon systems, apart from naval equipment, rather than producing them in small numbers. The current situation of the Danish arms industry reflects this centuries-long tradition.

In the aviation sector, Danish Aerospace is the Danish Air Force's main contractor for the maintenance, repair and modernization of aircraft and armament systems. It also produces electronics for aviation.

In the field of land systems, the company Hydrema, which offers the MCV-2 demining vehicle, should be mentioned.

Denmark has a number of powerful electronics companies, the most important of which is Teledyne-Reson as a supplier of sonar systems for combat ships.

4.5 Germany

History

German history books let German history begin with the Battle of the Teutoburg Forest in 9 AD. This is certainly not correct, because there were no Germans at that time. But since when are there really Germans and since when can one speak of Germany. The word "German" only appeared in the 8th century and it only denoted the language spoken in the eastern part of the Frankish Empire. The term "German" was increasingly transferred to the inhabitants of the areas that spoke this language. Since the 15th century one can therefore speak of a Germany. The transition from the Eastern Franconian to the German Empire is usually set with the year 911, in which after the Carolingians died out, the Franconian King Konrad I. (11-919) was elected king. He is considered the first German king. The authority of the king was not recognized from the outset without contradiction. Only if the king had military strength and pursued a skilful alliance policy could he gain respect among the powerful tribal princes. From Otto I (936-973) onwards, the German kings were also crowned emperors of the "Holy Roman Empire of the German Nation". In the course of the centuries the kingship passed from the Ottonians and Salians to the Staufer. With the fall of the Staufer in 1268, the universal occidental empire came to an end. In the late Middle Ages, Rudolf I (1273-1291) was the first Habsburg to take the throne. Since 1438 the crown had been de facto hereditary in the House of Habsburg. The appearance of Martin Luther and the spread of his Reformation set the entire social fabric in motion. In 1522 there was an uprising of the imperial knights and in 1525 the Peasant War, the largest revolutionary movement in Germany. Both uprisings were bloodily crushed. The aggravation of religious antagonisms finally led to the outbreak of the Thirty Years' War. Germany was devastated in this war and lost a third of its population. After the Peace of Westphalia in 1648, Germany was divided into almost 1700 countries. The individual countries adopted absolutism according to the French model. It gave the ruler unlimited power and opened the door to any arbitrariness, but also made it possible to establish a tight administration, introduce a well-ordered financial economy and raise standing armies. The economic policy of mercantilism also gave the states economic strength. Because of the many countries, there was no imperial patriotism in the territory of present-day Germany. The absolutist rulers were shaken to their foundations by the French Revolution, which swept over the territory of present-day Germany from the West. Emperor Napoleon also influenced the events in Germany. In 1806 the "Holy Roman Empire of the German Nation" came to an end. After the victory over Napoleon in the Congress of Vienna in 1814/15, many Germans hoped for a free, unified nation state. The German Confederation, which replaced the old Reichres, was a loose union of sovereign states. The only organ was the Bundestag in Frankfurt, not an elected parliament, but a congress of envoys. Just as the French Revolution of 1798 left its mark on Germany, the Revolution of 1848 also had a lasting effect on Germany. Popular uprisings took place in several countries, but they were all suppressed. In 1850, however, the German Confederation was restored. The following years were marked by economic upswing. Prussia became the hegemony in Germany. Under the Prime Minister Otto von Bismarck, Prussia consolidated its supremacy by winning the Danish-German War in 1864 and the war against Austria in 1866. As a result, Austria had to leave the German scene, the German Confederation was dissolved and the way was clear for Prussia to complete German unification. After the victory in the Franco-Prussian War in 1870/71, the southern German states joined together with the North German Confederation in Versailles to form the German Reich. Wilhelm I. of Prussia was proclaimed German Emperor. As a result, an economic upswing of undreamt-of magnitude began in the German Empire. The German Empire upgraded strongly, acquired some colonies and entered into an alliance with Austria-Hungary. When the heir to the Austrian throne was assassinated in Sarajevo on June 28, 1914, the German Empire entered World War I on the side of Austria-Hungary and, after an eventful war, emerged as the loser from the first worldwide trial of strength. The monarchy was subsequently abolished. In the Peace of Versailles the German Empire had to renounce the colonies and cede territories in Europe to its neighbouring states. In addition, it received conditions regarding the strength and equipment of its armed forces. The economic depression and the disgraceful peace treaty of Versailles were the breeding ground for the rise of the National Socialists, who from 1934 were given full power in the state. Since then, Germany has invested in armaments and infrastructure. Furthermore, they withdrew from the League of Nations and took part in the Spanish Civil War on Franco's side with the Legion Condor. On 1 September 1939 Hitler unleashed the Second World War by attacking Poland. After the lightning campaigns at the beginning of the war and the victories over France, the Balkans and Northern Europe, the beginning of the end began in 1941 with the attack on the Soviet Union. In May 1945, at the end of the war, large parts of Germany lay in ruins and millions of Germans had fallen or were in the camps of the Soviet Union. Germany was an occupied country after the war and it remained a divided country after the founding of the Federal Republic of Germany in 1949 and the German Democratic Republic in 1949. Both states were frontline states during the Cold War and accordingly highly equipped with their own strong arms industry.

Defence Industry

In the first half of the 19th century, Germany experienced faster industrial development than the then leading industrial countries England and France. The German Empire's share of the rapidly growing world industrial production doubled between 1820 and 1913. The relatively rapid industrial growth in Germany was mainly due to the consistent exploitation of the new technologies of chemistry and electricity and the transition from iron to steel as an industrial raw material. In addition, the division of labor, mechanization and concentration continued to advance. With its rapidly growing economic power, the German Empire demanded its share of the already divided world. This necessarily brought the German Empire into confrontation with the British Empire, as the latter was interested in maintaining the status quo. The German desire for a redistribution of the international balance of power led to the development of an armaments economy oriented primarily towards warshipbuilding. For England was above all a maritime power. Beneficiaries of the increased maritime armaments production were initially the shipyards, for example the Vulcan shipyard in Stettin, Blohm und Voss in Hamburg, the Germania shipyard and the Howaldtswerke in Kiel. By 1914, 38 liner ships, 13 large and 39 small cruisers, 150 torpedo boats and 44 submarines had been built at the German shipyards. In addition to shipbuilding, the financially strong large-scale industry also showed great interest in aircraft construction. Until the outbreak of the First World War, the number of aircraft manufacturers rose sharply. In addition to the Siemens-Schuckert factories, aircraft were built by the Allgemeine Elektrizitätsgesellschaft AEG, Gothaer Waggonfabrik, Flugzeugbau Friedrichshafen GmbH, Hansa- und Brandenburgische Flugzeugwerke AG, Hannoversche Flugzeugwerke GmbH, Rumpler Luftfahrtzeugbau GmbH, Deutsche Flugzeugwerke GmbH, Flugmaschinenwerke August Euler, Fokker Aeroplanbau GmbH, Albatros.

The most important manufacturers of pistols are Heckler and Koch as well as the Carl Walther arms factory, and for sniper rifles the world-famous Mauser company. Automatic rifles and submachine guns are manufactured by Heckler und Koch. Heckler und Koch is also known for extremely innovative solutions, for example an automatic rifle for caseless ammunition. The most important producer of machine guns is the company Rheinmetall DeTec.

The decimation of the peacetime strength of the German Bundeswehr primarily affects the construction of heavy weapons. In the past, German industry produced around 15,000 armoured vehicles after the end of the Second World War. Some 2,500 companies were involved in this process, manufacturing individual major assemblies and components on a division of labour. Most of these were medium-sized companies spread across the old federal states. While the sales volume in the 1980s was around 1.2 billion euros, the order volume fell to 250 million euros by the end of the 1990s. As the volume of orders fell, the number of employees in the tank construction sector also dropped from 55,000 in the mid-1980s to 15,000 in 1997. Since the Bundeswehr planning does not envisage any new large-scale production of combat vehicles within the armed forces, the entire industry is currently in a situation that threatens its very existence. The entire industry sector must therefore endeavour to obtain more export orders and maintenance contracts from the Bundeswehr. Opportunities for conversion in tank construction arise only in marginal areas, such as training and testing equipment, which can also be used for civilian production.

Krauss-Maffei-Wegmann, with operating locations in Kassel and Munich, is the leading system company in the field of armoured wheeled and tracked vehicles in Europe, thanks to the combination of complementary products and components. KMW develops and implements concepts for future light and heavy armoured vehicles in the areas of main battle tanks, artillery, air defence systems and pioneering equipment. With the latest developments, the PUMA infantry fighting vehicle, the Fennek reconnaissance vehicle and the Dingo all-protection vehicle, KMW is trying to regain market share on the world market. KMW has several subsidiaries that offer tailor-made logistics solutions and computers for artillery systems.

As a full-range supplier of land forces technology, the renowned defence contractor Rheinmetall-Detec develops and supplies a wide range of equipment, from various types of ammunition to artillery and main battle tanks as well as light and medium armoured combat and transport vehicles. The company equips naval forces with command and control systems and provides systems management for complex procurement projects. Rheinmetall provides the German Air Force with state-of-the-art flight simulators and on-board weapons. Air defence systems are distributed by the subsidiary Oerlikon Contraves.

As a producer of missiles and intelligent ammunition, Diehl has earned an excellent reputation worldwide.

After the end of the war, all the innovative achievements of the German aerospace industry were exploited by the victorious powers and it was not until the 1960s that Germany was able to devote itself again to the development of aerospace technology. The German aviation industry is highly concentrated. In terms of regional distribution, the focus is on Bavaria, followed by Baden-Württemberg, Hamburg, Lower Saxony and Bremen.

Probably the multinational armaments company Airbus, which through Eurofighter GmbH has developed the Eurofighter for Great Britain, Italy, Spain and Germany, which is also used by Austria, is currently making the most headlines in Europe. With the Eurofighter, Europe can come close to the technological lead of the USA in aviation products. In addition to the Eurofighter, Airbus is also involved in the development of Unmanned Combat Aerial Vehicles. Airbus is also involved in the modernization of existing weapon systems.

Another important area is the Airbus Helicopters division, which was created by the merger of the helicopter sectors of the German DaimlerChryslerAerospace and the French AerospatialeMatra. Since 2000, the Spanish company CASA has also been involved. Probably the most important product of this most successful European helicopter manufacturer is the Tiger multi-purpose combat helicopter.

After the destruction of the shipyards in the Second World War and dismantling by the victorious powers, the production of civilian ships was initially only restarted after the war. It was only after the establishment of the German Navy, which was initially equipped with US equipment, that the need for own naval shipyards gradually arose. The first warships from domestic production were delivered to the German Navy in 1961/62. The production is governed by the Paris Treaty of 1954, which prohibits the German Navy from owning nuclear submarines and large warships. These restrictions forced the shipyards to specialize in certain types of boats. It is in these niches that German warshipbuilding has had the greatest export successes to date, for example in minesweepers and minesweepers and conventionally powered submarines.

The shipyard in Emden belongs to ThyssenKrupp and together with Howaldtswerke is developing and producing the new Type 212 submarine for the German Navy. It also supplied Dolphin-class submarines to Israel.

The traditional shipbuilding company Blohm+Voss Hamburg belongs to Lürssen and produces MEKO frigates and for the German Federal Navy frigates of the F-124 class and develops corvettes of the K-130 type.

ADM Kiel's main production area includes all types of combat ships. Another producer of naval ships is Lürssen in Bremen. This shipyard specializes in speedboats, minesweepers, minesweepers and support ships.

Germany has a number of powerful companies in the electronics and communications industry, such as the globally successful Siemens Group.

Within Zeiss Optronik GmbH, the Observation and Target Systems business unit is responsible for the design, development and production of optical and optronic periscopes and platforms as well as paramilitary observation and vision systems.

Denmark has only limited capacity, but has built combat ships at Danish shipyards - the picture above shows a fleet supplier.

The German defence industry produces a wide range of military equipment:

Above: Tiger combat helicopter, Dingo armoured all-purpose vehicle

Middle: Transport planes, battle tanks, bottom: Submarine

4.6. Finland

History

The geographical location of the country has repeatedly made Finland a scene of neighbourly disputes. From the 12th century Finland became a Swedish province. Despite Swedish immigration, the Finns retained their ethnicity. Although the Swedes lost parts of Finland to Russia in the Nordic War, Finland only came completely into Russian hands when Tsar Alexander I annexed the country without declaring war in 1808 and declared it an autonomous Russian Grand Duchy. After the Russian October Revolution in 1917, the parliament declared Finland independent on 6 December 1917. After a brief war of independence, which General Mannerheim won with German help, Finland became a republic in 1919. The rejection of the Soviet demand for military bases on Finnish territory in 1939 led to the Winter War. In 1941, Finland fought on the German side against the Soviet Union in order to regain the territories lost in this war, but in 1944 it concluded an armistice with the Soviet Union, which saved it from Soviet occupation. In the peace of 1947 Finland had to renounce Petsamo. Finland is neutral but is a member of the European Union.

Defence Industry

The Finnish defence industry does not have the centuries-long tradition as for example the defence industry in Sweden or in other similarly large countries in Europe. The foundation stone of an independent arms industry was only laid with independence from Russia from 1917 onwards. From this time until the end of the Second World War, the industry produced small arms, grenade launchers, motor vehicles and also aircraft. The Finnish defence industry currently has almost 50 companies engaged in the development and production of military equipment. The most important defence company in Finland is the Patria Group, founded in 1997. Since 2004, the Group has consisted of four business units: Patria Vehicles, Patria

Weapon Systems, Patria Aviation, Patria System Solutions. The Vehicle Division develops and produces armoured vehicles and also provides lifetime service. The most important export hit is the Patria XA-200, which is particularly well proven in UN peacekeeping operations. In addition to its own development, the company also produces the CV 9030 Alvis Hägglund main battle tank under license for the Finnish armed forces. In the Weapon Systems division Patria develops and produces state-of-the-art grenade launcher systems and field guns. In addition, the company also deals with the modernization of existing weapon systems. The most recent major success to date is the AMOS (Advanced Mortar System) grenade launcher system, a grenade launcher system developed jointly with Alvis Hägglung which can be used both on land and at sea. The Aviation Division is engaged in the maintenance of Finnish Air Force combat aircraft, including their modernization, and is also the Northern European competence center for the NH 90 European helicopter. The System Solutions Division in turn deals with the contract-based development of solutions for special requirements, such as night vision technology and the development of service facilities for airports for specific weather conditions.

Finland has a number of shipbuilding companies. The most important have sites in Helsinki, Rauma and Turku and develop and produce civilian cruise ships and ferries as well as warships and icebreakers. The shipyards developed the Hamina class, a rocket-powered speedboat with a displacement of around 236 GRT and a speed of more than 30 knots, for the Finnish Navy, among others.

The main ammunition manufacturer in Finland is Nammo, a company founded in 1923 as Lapua Ltd. As a result of the Nordic Ammunition Group founded by the three Northern European countries in 1998, Lapua became part of Nammo. The company offers a wide range of ammunition types.

Finland also has a producer of small arms. The Riihimäki-based company Sako Oy used to produce automatic rifles and a light machine gun for the Finnish Army. At present, the company has a high-performance sniper rifle in its product range.

Finland also has a large number of electronics companies producing goods for the military.

4.7. France

History

The territory of present-day France came into the light of history for the first time when Julius Caesar invaded Gaul and conquered the land inhabited by Celts for the Roman Empire between 58 and 51 BC. After the end of the Western Roman Empire, Germanic tribes invaded the country and founded new empires. The French like to begin their history with Charlemagne (768-814). The following centuries bring with them an eventful history with different systems of rule. Through a marriage of the fiery French noblewoman Eleonore of Aquitaine to the English King Henry Plantagenet in 1154, almost half of France came under English rule. English claims to the French throne lead to the Hundred Years' War (1339-1453), which Jeanne d'Arc brings about the turning point in 1429. With her, a completely new force enters the war: a French national feeling, the term "la patrie", is created. With the expulsion of the Englishmen-who could only hold Calais (until 1558) and the Channel Islands-strengthens the kingship. Under Louis XI, France finally becomes a nation state. The further rise of France to a great power is first marked by the participation in wars in Italy and Germany. France's unscrupulous rise to a great power seems unstoppable when in 1562 the self-destruction by the Huguenot wars begins. With the accession of the Bourbons, a new era began in France. Louis XIII (1640-1643) was a ruler who suffered greatly under his minister Cardinal Richelieu. The climax of the power development of the Bourbons takes place under Louis XIV (1643-1715), who sees himself as the "Sun King". Under Louis XIV and his minister Colbert, the country achieved wealth through mercantilism. In order to increase the country and fame further, Louis XIV starts robbery wars against Habsburg. The robbery wars bring small land gains, but enormous blood losses. Louis XIV sits on the throne for 62 years; his great-grandson Louis XV (1715-1774) reaches 59 years. Under his reign, the splendour of the god-kingdom slowly breaks down due to mismanagement, immorality and wars. During the reign of Louis XV, France lost its overseas colonies, but won Lorraine and Corsica. Under his successor, Louis XVI (1774-1792) it was already too late for reforms. Dammed up counterforces finally discharged into the French Revolution. This revolution ended in a dictatorship, but by abolishing the feudal system and absolutism it created the conditions for a more modern age. With the fall of the Board of Directors in 1799, Napoleon Bonaparte became the first consul to head the state. The coronation as emperor Napoleon I. takes place in 1804. Napoleon was on the road to victory for a long time, until his power came to an end after a failed Russian campaign in 1812 and defeat at Waterloo in 1815. At the Congress of Vienna, France was obliged to return all territories acquired after 1792. France underwent restoration under the Bourbons. The period between 1815 and 1948 is marked by several riots in which the people want to enforce more rights. In 1848, as in many places in Europe, there was a revolution in France. The king had to abdicate and the second republic was declared. A nephew of Napoleon I is elected President of the Republic, but after a coup in 1852 he is elected Emperor. As Napoleon III he leads his country into the Crimean War (1854-1856), supports Italy against Austria in the war of 1859, but finally loses the Franco-Prussian War of 1870/71. After the defeat the Third Republic begins. The period of the Third Republic is characterized by the expansion of the colonial empire and by participation in two world wars. France emerged victorious from both world wars, but immediately after the war the economy was severely disrupted and disrupted. With the constitution of October 1946 the Fourth Republic came into being. It changed from a French colonial empire to the French Union. In the 12 years of its existence it went through severe internal and external crises. Inside, the communists became stronger and stronger and the reckoning with the former collaborators with Nazi Germany weighed heavily on the climate. In terms of foreign policy, France could not defend itself against the separation of the colonies. In 1954 Indochina was lost after heavy fights, in 1956 Morocco, Tunisia and also in Algeria there were already uprisings which aimed at a separation from France. Based on the negative events of the Fourth Republic, Charles de Gaulle undertook the drafting of a presidential constitution tailored to his person, which was approved by a referendum. De Gaulle was subsequently elected President of the Republic. Parliament's power was severely restricted. Once internal affairs had been settled, de Gaulle set about repositioning France in the world. He detached the country from the military organisation of NATO and built up a French nuclear force of its own. De Gaulle handed over power to his successor Pompidou in 1969 and died the following year. The biggest upheaval in the history of the fourth republic was the victory of the socialist Mitterand in the 1981 presidential election.

French armaments industry

The first documented French mention of artillery dates from 1338, in a document referring to the arrangements made in the Rouen arsenal for the dispatch of military equipment for the ships moored in Harfleur to be equipped for an attack against Southampton. It mentions an iron cannon with 48 feathered iron bolts, a pound of saltpetre and half a pound of sulphur to drive the arrows.

The most important producer of artillery guns in France at the end of the 19th century was the Schneider armourer.

In contrast to Germany, France, at the beginning of the First World War, recognized the usefulness of armored land vehicles which, armed with a cannon, can knock down machine gun nests. In May 1915, Schneider, the largest French armaments company, ordered two Holt-Caterpillars from the United States of America and undertook trials with them. In mid-December, Schneider invited the head of the French Ministry of Invention. A small 45 hp Holt tractor with a box-shaped body was presented to him. The vehicle was developed by Schneider's chief designer, Eugene Brillie. It was underpowered, but it met with the approval of the civil servant. After the French commander-in-chief General Joseph Joffre had also given a positive report, Brillie started work on the Schneider Kampfwagen in December. The work progressed so quickly that Joffre was able to demonstrate the first prototype of a combat vehicle based on the Holt tractor as early as February 1916. After the demonstration, a contract was signed for the production of 400 Schneider CA-1 main battle tanks. But it should still take a year that enough vehicles for the front line were available. The tank weighed 14.6 tons, had a 7.5 cm cannon and two 8 mm machine guns as armament. The crew was six men. The tank experienced its first deployment on 16 April 1917. 57 of the 132 armoured vehicles used at that time were immediately destroyed and many others were rendered useless, so that repairs were no longer worthwhile. However, the Schneider combat vehicle already had competition before it was even produced. It was the Char d'Assaut of the Compagnie des Forges et Acieries de la Marine et d'Homecourt in Saint Chamond, which was completed as a prototype as early as 1916. But just like the Schneider battle tank it took almost a year until the tank became effective on the battlefield. The tank was built at FCM La Seyne near Toulon. The Chamond weighed 22 tons, had a 7.5 cm cannon as main armament and four machine guns as secondary armament. The crew was 9 men. The first operation took place on 5 May 1917. The two combat vehicles proved to be too cumbersome, so Renault also started to construct a battle tank. Renault's combat vehicles, called FT-17, were first used in combat on 31 May 1918. The tank weighing about 6.8 tons with a 3.7 cm cannon as main armament turned out to be a great success and was used not only by France but also by Belgium, Japan, Brazil, Canada, China, Czechoslovakia, Finland, Greece, Italy, Manchuria, the Netherlands, Poland, Spain, Great Britain, USSR, Yugoslavia and the USA. A little more than 3,700 units of type FT 17 were produced in several variants.

In 1921 France developed its first tank program. In this program they defined two types of tanks. A char de rupture (tank for breakthrough) and a char de bataille (battle tank). One char de rupture and five types of the char de bataille were built under the leadership of the section "battle tanks", whose head was the tank inspector General Estienne. The model of a char de rupture goes back to the first missions of tanks during the First World War, when, according to the findings of the mission of the three French tanks, a heavy battle tank was required, which should have a rhomboid-shaped running gear. In addition, it was to have armour plating that would make it strong enough to be invulnerable to a breakthrough. The first prototype was a 40 tonne tank, the Char 1A. The end of the war prevented a final production. Instead, a heavy battle tank based on the Char 1A was designed and developed until it was ready for serial production. The new heavy battle tank was named Char 2 C and weighed 70 tons. The crew was 12-13 men, the armour was 13 to 45 mm thick. The tank was armed with a 7.5 cm cannon and four 8 mm machine guns. The ammunition equipment was 125 shells for the cannon and 10,000 cartridges for the machine guns. It was planned to build 300 pieces of the Char 2 C, but in total only 10 pieces were produced. They were still in use when the German troops marched in 1940. Regarding the main battle tank there were five proposals for the 1921 program, but it was not until 1927 that the first prototypes of the Char B were ordered from Delauney-Belleville, FAMH (St. Chamond), FCM and Renault/Schneider.

Required was a 15 tonne tank and a 4.7 or 7.5 cm cannon mounted on the hull. When the first prototypes appeared between 1929 and 1931, they weighed about 25 tons. Based on experience in manoeuvring, further demands were made, which led to the production of a modified version of the Char B as Char B 1. Only 36 of the Char B 1, which joined the troop in 1935, were produced and the order for a further improved version Char B 1-bis was placed. The Char B 1 to weighed 32 tons. Renault/Schneider produced about 365 Char B 1 to Char B 1 to until 1940.

In 1926 the first tank construction program was replaced by a new program. The new program saw the production of a char leger (light tank) with a 3.7 or 7.5 cm cannon, a char de bataille (main battle tank) with a 4.7 cm or larger cannon and a char lourd (heavy main battle tank) with strong armor. In 1926 there was an urgent need to replace the aging Renault FT-17. But the tight budget was only enough to build a modified version of the FT-17 which was produced as Char NC1/NC27. Neither type was accepted by the French army, but it was exported in small quantities. Other char legers were the Renault R-35 and R-40, the FCM-36 and the Hotchkiss H-35. All these tanks weighed around 10 to 12 tons. The Hotchkiss H-35 and its successors H-39 and H-40 were produced until 1940 in about 1100 pieces. About 1900 pieces of the Renault R-35 and R-40 were produced. Of the FCM-36, which had a diesel engine, only 100 were produced until 1940. The next, stronger tank type were the chars moyen (medium battle tank). In this category can be classified: Renault D 1 and D 2 and the AMX-38. 14 tons of the Renault D1 and D 2 weighed 14 tons and were produced in 210 pieces. Of the AMX-38 was a 16 ton tank. A particularly successful design was the Somua S-35, which was produced by the Societe de'Outillage Mecanique et d'Usinage d'artillerie (SOMUA). The armour consisted of a hull made of cast steel and not of armour plates as was previously the case. The Somua weighed 20 tons and had a 4.7 cm cannon. In 1940 an improved version with a more powerful engine appeared on the battlefield, the Somua S-40. 500 S-35 and S-40 were produced by 1940. A special type of tanks were the Auto-Mitrailleuse de Reconnaissance (AMR), light tanks with a combat weight of about 5 to 10 tons. About 200 pieces of the AMR were produced.

After the liberation of France after the landing in Normandy in 1944 the construction of new tanks was started immediately. The first new constructions were a medium battle tank and a fighter tank. The AMX-13 weighed about 15 tons, was lightly armoured and had either a 7.5 cm, 9 cm or 10.5 cm cannon. Its favourable price made it an export hit after 1953, the year of its introduction. The ARL-44 main battle tank was a 48 tonne combat vehicle with a 9 cm cannon. Only 60 units were built between 1947 and 1953. Immediately after the end of the war, France developed a heavy battle tank, the AMX-50, which never went into series production. The reason for this was the aid program of the USA. For France the M 47 was available free of charge through this program.

After the joint project between Germany and France to build a main battle tank had failed in the mid-1950s, France designed a new medium battle tank, the AMX-30, which went into series production in 1960. The AMX-30 weighed 37 tons and had a 10.5 cm cannon. GIAT produced round pieces of the AMX-30. It was also exported to many countries. The next tank developments were the AMX-32 and the AMX-40. Both battle tanks were further developments of the AMX-30 and had better armament, a 12 cm cannon and stronger armour. Much better than the AMX-30, 32 and 40 is the Leclerc. Its development dates back to 1983, after the second attempt of a joint German-French tank development had failed. The Leclerc with its 54 tons combat weight is a state-of-the-art main battle tank.

The beginning of the history of the French fleet can be traced back to the reign of Philip II. August (1180-1223). The conflicts that began with England at that time gave rise to a series of maritime ventures, in which, however, fleets formed only ad hoc were used. One example is the fleet of 80 large and almost 600 smaller ships assembled by Eustach the Monk in Calais in 1216, with which the Dauphin Louis wanted to carry out an invasion of England. However, the fleet was scattered by a storm, so that the Dauphin only arrived at the Thames with a small number of ships. Eustach then assembled a further fleet of 80 ships with which he wanted to come to the aid of the dauphin. But he was defeated by the English at Dover in 1217. Eustach was killed in this naval battle and the Dauphin had to break off the enterprise. With only 15 ships he reached the port of Calais. Under the name of Louis IX (1226-1270) he created the title of "Admiral of France" for the commander of a French fleet. The next king who again had a fleet equipped was King Philip IV. (1285-1314). With a fleet of about 30 larger ships, he achieved a naval victory over the Count of Flanders. The next epoch, which also included sea battles, was the Hundred Years' War between France and England, which began around 1336. During the war, a French fleet suffered a crushing defeat at Sluys in 1346. In 1374, Admiral Jean de Vienne tried to create a royal fleet at Rouen. He had ships built and only three years later he had a fleet of 35 large and well armed ships. But the foundation of a fleet was only of short duration. It was not until the 16th century that the interest in the voyage to the New World to Newfoundland began. One took these trips as an opportunity to capture Spanish and Portuguese ships. The French King Francis I (1515-1547) issued letters of marquees for the purpose of fighting against Portuguese and Spanish ships. However, all the efforts to establish a permanent French fleet were short-lived. It was to be Jean Armand Duplessis, better known as Duke of Richelieu, who directed history as a senior minister between 1624 and 1642 under King Louis XIII (1601-1643). Richelieu promoted shipbuilding and increased the staff of naval officers. At his death, the French fleet comprised over 30 large and 27 small warships. In terms of shipbuilding, the ships could be described as modern. The successor of Richelieu, Mazarin (1643-1661) had no interest in a fleet. Only the successor Colbert (1661-1683) continued Richelieu's naval tradition. First Colbert had to carry out his work against the will of King Louis XIV (1638 -1715). Colbert proved to be a tireless worker and excellent organizer. Under his supervision ports and shipyards were soon established. Brest was extended to a sea fortress. At Colbert's death the French fleet comprised 150 large units and 30 galleys served as coastguards. During the reign of Louis XIV, two outstanding fleet leaders, Abraham Duquesne and the Count of Tourville, achieved extraordinary sea victories over the Dutch and English. Another naval hero who soon gained popularity through his unusual sea battles was Jean Bart. After the era of the three great naval leaders Duquesne, Tourville and Jean Bart, a new era of French seafaring began with the War of the Spanish Succession. The French fleet was financially starved after Colbert and therefore had to reduce the number of liner ships from 120 to 85. Thus the French fleet entered the War of Spanish Succession against the English navy, weakened. During the war France lost 40 liners and 150 smaller warships, which could not be replaced due to lack of money after the War of Spanish Succession. This fact increased the predominance of England at sea. The neglect of the navy was to take bitter revenge on the securing of colonial conquests. While England was able to protect its colonies with a large fleet, France was not able to protect its merchant fleet. France was as much involved in the next wars as England. France fought a duel with the English fleet during the Seven Years War and during the War of Independence of the English colonies in North America. France lost almost all its maritime disputes with England, only when the balance of power developed in favour of France did the naval leaders de Grasse and Suffren win. During the French Revolution, the fleet suffered great defeats, so that under Napoleon I, immediately after the Egyptian adventure, the fleet reached a low of 51 ships. During the time of Napoleon I, the French fleet did not achieve great victories. On the contrary, the naval battles of Aboukir in 1798 (France lost 11 ships of the line and two frigates and 1000 men dead or wounded out of a total of 13 ships of the line and 4 frigates) and Trafalgar in 1805 (the French and their allies lost 18 ships and 2600 men dead or wounded out of a total of 33 ships of the line and five frigates), losing ships and sailors that France could never replace. France then practically gave up naval warfare and maritime trade also came to a standstill. Likewise, France had no significant colonies from about 1810 onwards. Although 80 liner ships were built after Trafalgar, there was no possibility to train the crews and to accustom them to the war.

The establishment of a French fleet after the Napoleonic era was initially preceded by a wave of purges among the naval officers. The first newbuildings were not started until 1821, but only very moderately, e.g. ships were lying half-finished for years as reserve in the shipyard. It was not until 1846 that Admiral Mackau persuaded the Chambers to adopt a somewhat ambitious shipbuilding programme, but its execution was soon halted by the financial difficulties of the Second Republic. At that time, technical innovations and a replacement of the old cannons took place. The colonial policy that began in 1827 contributed to a renewed focus on naval supremacy. Between 1830 and 1840, the number of personnel was increased and training improved. Since 1837 there was also an African squadron, which was commanded by Rear Admiral Lalande. The squadron consisted of six liner ships, which enjoyed great respect even among the English. As a "squadron in being" it was an instrument of politics. In the 1940s, a confrontation with the caper war began. Admiral Mackau succeeded in convincing the chambers that France needed a strong fleet again. In 1846, the Chambers finally determined that France should have a fleet of 44 large liner ships, 66 frigates and 135 auxiliary ships. An acceleration of warship building was brought about by the threat of the Crimean War. In the run-up, everything was done in the shipyards of Toulon in order to make the ships ready for sea. When France entered the Crimean War, it had a fleet of 300 war ships, thereunder 100 steam ships. Napoleon III. was a promoter of the navy. He recognized very early the connection between propeller, heavy guns and tanks. Therefore in France, three ships were first stacked on piles, which were constructed according to this connection. Although these ships were not able to sail to the location in the Black Sea under their own power, they forced the decision before Sevastopol with their heavy artillery. The next step was to construct a ship that was able to sail with its own propulsion despite good armour. The result was the "Gloire", which was started in 1857 in the shipyard Toulon, launched in 1859 and reinforced in 1860, and which made all wooden ships of the line obsolete with one blow - comparable to the H.M.S Dreadnought half a century later. The Gloire had a displacement of 5630 tons, was 77.88 m long-16.99 m wide and a draught of 8.48 m. The armament consisted of 36-16 cm guns M1858/60 and the armour was 4.7inch armour plates over the wooden hull. The steam engine gave it a speed of 13 knots. The crew was 570 men. The sister ship Invincible was also built in Toulon. The third ship of the type, the Normandy, was built in Cherbourg. The development of the French Navy during the time of Napoleon III was closely related to the colonial policy of the emperor. Since 1854, France worked on the colonial empire and incorporated piece by piece land in Africa and Asia. Around 1870, the French fleet already had 19 armoured ships in its arsenal. When France lost the war against Prussia, the navy emerged from the war without material losses or loss of prestige. However, the defeat in the war led to a reduction of the fleet building programme and a discussion arose in democratic, republican France about the value of battleships and the possibilities of naval wars at the end of the 19th century. In this context a "Jeune Ecole-Junge Schule" was formed, with Richild Grivel as its earliest representative. He wrote his book "La Guerre des Cotes" in 1864. According to his ideas, the weaker France could only wage a cruiser war against England - with more psychological effect on trade and population than with remarkable successes against the English fleet. Soon after the publication of the ideas of the Young School, the Republicans became enthusiastic about the ideas. A mutual community of interests with the left Republicans, radical Socialists and a part of the Socialists began to develop. In the middle of the 1880s, the followers of the Young School found themselves in the frequently quoted torpedo frenzy of 1884, after, for example, Etienne Lamy had summarily described the battleship politically as a prestige unit of the Empire, an aristocratic liner, in 1882 and polemically contrasted it with the democratic torpedo boat, which brought the advantages of equity and the many commander posts. Under the pressure of such political arguments, the chambers rejected the construction of further battleships in 1881: From 1882 to 1889 no more large battleships were stacked and the completion of the ships under construction was repeatedly delayed. In the meantime, the torpedo boat was increasingly upgraded in terms of combat value and operational seaworthiness. It was not until the naval manoeuvres of 1886/87 that the ideas of the Young School lost credit, as the torpedo boats suffered one failure after another against the battleships. So the ideas of the Young School came to an end around 1887 and the Old School regained its position without the ideas of the Young School having been carefully examined for at least a few years. During the manoeuvres of the following years only the ideas of the old school were rehearsed. Thus the battleships became bigger and bigger and therefore more expensive, while the development of the torpedo carriers was consciously postponed. The theory of the Young School could only hope for a renaissance again in 1902 after the Socialist Camille Pelletan took office as Navy Minister. In 1902, he introduced a new programme that envisaged the construction of fast ships, flotillas and bases. At the same time, he requested an indefinite postponement of the construction of liner ships and in 1904 he drew up a construction programme which from 1905 onwards envisaged 14 large ships, but also 168 torpedo and submarines. However, the end of the renaissance of the Young School came from the Far East, when the Imperial Japanese Navy had crushed the Russian Navy in the Naval Battle of Tsushima. The Young School disbanded. What remained was that around 1905 the French Navy was a patchwork of different ideas and the floating arsenal was at a material low. Pelletan resigned in January 1905. In 1905, the new Minister of the Navy, Rouvier G. Thomson, published a kind of interim draft of a new naval construction programme. This planned a fleet of 34 liner ships, 36 armoured cruisers, 279 torpedo boats and torpedo boat hunters as well as 131 submarines. But because the chambers wanted the construction of Dreadnough, the program was not implemented. Instead, a new fleet law was passed in 1912, which provided for the construction of 28 liners, 10 light cruisers, 52 squadron torpedo boats, 94 submarines and 10 large units for foreign service. In 1919 the fleet was to be ready for service. At the beginning of the First World War, France had 4 large line ships, 18 liner ships, 18 armoured cruisers, 9 cruisers and 80 destroyers and large torpedo boats and 16 submarines. During the war, the French Navy unobtrusively performed considerable services and suffered heavy losses. During the First World War, France lost about 113,000 tons of warship space. At the end of the war, the French fleet ranked behind the Italian and Japanese and claimed a part of the German fleet. However, the USA and England had already decided to sink the German fleet before the end of the war. After the war, the construction of submarines was increasingly turned to. After the war, the total amount of warshipbuilding at French shipyards after the war was only about 7%, as all shipyard capacities were needed for the construction of merchant ships. A further restriction of warship building was brought about by the Washington Naval Conference, at which the quantitative strength ratio of the five naval powers USA, Great Britain, Japan, Italy and France was set at 5:5:3:1.75:1.75. Under these conditions France was able to build 175,000 tons of battleships, 60,000 tons of aircraft carriers and an unlimited number of cruisers, destroyers and submarines. At the conference, France had to accept equality with Italy, a humiliating regulation that seriously injured national pride. After the Washington Agreement, France adopted a minimal programme: 8 battleships, 16 light cruisers, 36 large destroyers, 60 destroyers, 2 aircraft ships, 78 submarines and 8 submarine cruisers. The modern new ships were not built until the beginning of the 1930s, when it became clear that Germany was breaking away from all the armament restrictions imposed by the Treaty of Versailles.

In 1939 France had seven battleships, an aircraft carrier, seven heavy cruisers, 12 light cruisers, 58 destroyers and 77 submarines.

A.C. de la Loire, St-Nazaire, Arsenal Brest, Arsenal Lorient, C de la Gironde Bordeaux, F C de la Mediterranee La Seyne, Arsenal de Toloun, Penhoet St Nazaire, Dyle et Bacalan Bordeaux, Arsenal de Rochefort, Normand Le Havre, Schneider Chalon-sur-Saone, A et Ch de Bretagne Nantes

France was without doubt the number one air power in the early years. France already had its own air force before the First World War. These had 600 land and 20 seaplanes and a staff of 1900 men. Likewise, France already had an efficient aviation industry at the outbreak of the First World War. The most important entrepreneurs who produced aircraft were Bleriot, Brequet, Caudron, Dorand, Farman, Hanriot, Morane-Saulnier, Nieuport, Salmson, Spad, Voisin. It is worthwhile to take a closer look at the companies. Spad was founded in 1910 by Armand Deperdussin as Societe pour Avoins Deperdussin in Betheney near Reims. Deperdussin went bankrupt in 1913 due to embezzlement and the factory was taken over by Louis Bleriot in 1914. Bleriot renamed the factory Societe pour l'Aviation et ses Derives. After the war, Spad was integrated into Bleriot Aeronautiques. Spad was an extremely successful producer of fighter planes. They produced 8,472 Spad 13 and 6,000 Spad 7. Caudron was founded by Gaston and Rene Caudron in

1912 in Rue and produced for example 2,850 units of the bomber type G3, 1,421 units of the G4 and 310 units of the type R11. Hanriot was founded in 1909 by Rene Hanriot in Bezeny near Reims. Hanriot produced, among other things, 1,145 single-seater fighters of the type HD 1. Farman was founded by the brothers Dick, Maurice and Henry Farman in 1908 in Bouy pres de Chalous-sur Marne. Maurice and Henry founded the Societe Henry et Maurice Farman in Billancourt in 1912. Although the two brothers worked closely together, the two brothers gave their constructions their own names. On the other hand, however, the business cooperation became so successful that the joint factory was the largest aircraft factory in the world at the beginning of 1914. When, in their opinion, the war became inevitable, the Farman brothers started mass production. When war broke out, they were therefore the only company able to take on large orders. This also explains why, at the beginning of the war, the majority of the French Air Force consisted of Farman aircraft. Farman successfully produced large numbers of grid tails, including the F 40 model, a reconnaissance plane and night bombers. Morane-Saulnier goes back to the foundation of the company by Robert and Leon Morane and Raymond Saulnier in 1910. Their monoplanes, called "Parasol" because of the high-deck construction method, quickly became popular. The company produced about 1,080 MS-406 type monoplanes. Voisin was founded by Gabriel Voisin. The company mainly built bombers, including type VIII, around 1,100 units. Nieuport was founded as Societe Anonyme des Etablissements Nieuport in 1909 by Eduard Nie Port in Issy-les-Moulineaux. Chief designer of Nieuport was Gustave Delage. At the beginning of 1916, Delage achieved the biggest success with the construction of the Nieuport 17, a hunting biplane with excellent flight characteristics. Of the Nieuport 17 were round. The company Salmson, producers of aircraft engines were Renault, Gnome, Le-Rhone, Canton-Unne (Salmson) and Hispano-Suiza.

In the interwar period, France was unable to match the excellent performance of the aircraft manufacturers before and during the First World War, although new companies were created, such as Bloch (1930 by Marcel Bloch in Villacoublay), S.A.B. (1924 in Bordeaux), Latecoere (1922 by Pierre Latecoere in Montoudran) and Potez (1919 in Auberville by Henry Potez). Military aviation in France was therefore in a poor state in the mid-1930s. When the Second World War broke out, many fighter and bomber squadrons were therefore equipped with obsolete equipment. The bomber squadrons were equipped with obsolete Amiot 143 and Bloch 210's. The Amiot 143 M bomber and reconnaissance aircraft was first delivered to the Armee de l'Air in 1935 and was intended to be used as multi-seater fighters, day and night bombers and long-range reconnaissance aircraft. The aircraft was obsolete at the beginning of the Second World War. The more modern types Bloch 174 and Liore et Olivier LeO 451 were only available as prototypes at the time of the German attack and many aircraft of the type Potez 630 had to remain on the ground due to lack of propellers. Only the Breguet 693 proved itself in battle.

The most important and best fighter of the French Air Force was the Dewoitine 520, of which about 600 were built. The most important fighter at the outbreak of war was the Morane-Saulnier 406, of which about 1,040 were built until 1940. Further aircraft were the fighter Caudron 714, the transport and training aircraft Caudron Goeland, the obsolete bomber Farman 222, the fighter bomber Liore-Nieuport LN-40.

The first two helicopters that ever flew took to the air in France in the second half of 1907. One attempt was made by the brothers Louis and Jacques Breguet, the other by Paul Cornu. Another Frenchman was Etienne Oehmichen, who took his helicopter in 1924 on a sightseeing flight of one kilometre over 14 minutes. Louis Brequet gave up the development of a helicopter in 1909 and founded an aircraft factory. Nevertheless he continued to work on the problem of the helicopter. In 1929 he patented ideas for improving the stability of rotor aircraft. Two years later, he and his technical director Rene Dorand started building the "Gyroplane Laboratoire". This machine is generally regarded as the first truly successful helicopter in history. The first test flight took place in Villacoubly in 1935. By November 1936, the Gyroplane Laboratoire had set new world records with a speed of 98 km/h, an altitude of 158 m, a distance of 44 km and a flight time of over 62 minutes. Brequet's work was interrupted by the Second World War. After the Second World War, France made the greatest efforts to build helicopters. Although they had little experience with this type of aircraft, Breguet built a prototype of the "Gyroplane IIF" with coaxial rotors in 1949. However, the prototype never went into series production. Unlike Breguet, the state-owned company SNCASO (Sud-Quest) managed to develop a helicopter with the type designation "Ariel I". The Ariel I was followed by the Ariel II and the Ariel III. Another state group, the SNCASE (Sud-Est). SNCASE started their work in La Courneuve in the north of Paris by flying a Focke-Achgelis Fa 223 and, after some improvements, named it SE 3000. A further development, the SE 3130 was equipped with a Turbomeca Artouste turboshaft engine instead of the star piston engine. The machine was called "Alouette II". The Alouette II became the first series helicopter with a turbine engine. The success story of French helicopter developments began with the Alouette II. Production of the Alouette II ended in 1975, after more than 1300 units had been built and delivered to customers in 46 countries. Another highlight of French helicopter development was the Alouette III. The Alouette III first flew in 1959. The seven-seater helicopter proved to be an even greater sales success than the Alouette II. In total, up to one piece was produced and sold to more than 70 countries.

France went into the Second World War without much preparation of motor vehicles for military use. So France entered the war with a fleet of vehicles that was completely obsolete. Only in 1936 the Minister of Defence at that time ordered an armament programme. In the course of this program the armed forces received 3,200 modern tanks and 5,000 smaller tracked vehicles. The staff vehicles were mainly civilian vehicles built by Renault, Citroen, Hotchkiss. Real four-wheel drive military vehicles were produced by Renault, Berliet, Laffly in Asnieres near Paris, Latil in Suresne and Lorraine in Argenteuil. In France, all well-known vehicle manufacturers also produced trucks, such as Citroen, Renault, Berliet, Panhard, Willeme, Bernard, Matford and Peugeot. After the defeat of France the vehicle manufacturers had to produce for the German Wehrmacht. Thus Peugeot produced about 48,000 DK5 trucks and Citroen 15,000 23R models for the German Wehrmacht. The French vehicle industry also produced tractors for the artillery and half-track vehicles. During the First World War Latil, Panhard and Renault produced tractors. At the end of the war, about 2,000 Latil and 700 Renault were in service. In the interwar period Latil produced two light tractors for the artillery. After the defeat in the Second World War, the Latil factory came under the control of Daimler Benz. After that, tractor units were produced for the German Wehrmacht. Another company that produced tractors was Lorraine. Thanks to Adolphe Kegresse, France was a leader in the development of half-track vehicles in the interwar period. The designs of Kegresse were first built by Citroen. When Citroen was taken over by the tyre manufacturer Michelin in 1934, Kegresse founded its own company S.E.K. (la Societe d'Exploitation Kegresse). Production of the new Unic Type P107 construction began in 1938 at Unic in Puteaux. At the time of the defeat in 1940, 3,276 vehicles had been completed. Of another model, the Unic TU 1, only 236 were produced before the fall of France. After the occupation by the German Wehrmacht, production was carried out for the German army. Another producer of half-track vehicles was Somua, a company of the Schneider Group, which primarily built buses and trucks. The company produced several types of half-track vehicles.

The main companies in the French defence industry are then presented separately for land systems, aeronautics and naval shipbuilding, electronics and others, and finally the future of the French defence industry is forecast.

At the beginning of 2020, the French defence industry presents itself as an industrial sector consisting of almost 5,000 companies employing 165,000 people and employing a further 240,000 people through subcontracting. The industry as a whole represents about a quarter of the total European defence industry.

The most important entrepreneur involved in the development and production of combat systems for the land forces is Nexter, which emerged from the state-owned Giat Industries in 2006. The company's traditional business activities focus on wheeled and tracked vehicles, as well as small arms and ammunition. Nexter is also a supplier for shipyards and the aviation industry. With around 3,350 employees, Nexter generated sales of around €1 billion in 2014 in the business areas of armored combat vehicles, artillery systems, the automatic standard rifle Famas of the French armed forces, large- to medium-caliber ammunition, tactical communication systems for combat vehicles and comprehensive logistics support systems. Worldwide, some 18,000 armoured combat vehicles, 1,000 artillery systems, 13,500 medium-calibre weapons produced by Nexter are used in around 100 countries. Nexter pays special attention to the research and development of new weapon systems and the modernization of existing systems, for which it uses around 20% of its profits. In addition to research and development, the company's philosophy also includes numerous contacts with other defence companies, such as Bofors, United Defence and Renault. Nexter's best-known products include the Leclerc main battle tank, the light tank family around the AMX-10 and the Caesar artillery system, a 155mm artillery gun on a 6x6 vehicle with high mobility.

Another company that produces land systems is Renault Trucks Defense, which produces light tanks. The French Armed Forces are also a major customer of the French automotive industry, which supplies large numbers of off-road trucks and command vehicles.

The largest naval shipbuilding company is the **Naval Group** (until June 2017 *Direction des Constructions Navales (DCNS)*). The Naval Group has 13,600 employees and large shipyard facilities in Toulon, Brest and Cherbourg. The company is one of the largest shipyards offering complete solutions for naval warfare. Its customers include more than 40 countries around the world, including the USA, Taiwan, Saudi Arabia, Brazil and Italy. The shipyard can look back on a long tradition, as Cardinal Richelieu already had a state shipyard built in Brest in 1631, which was later expanded by Colbert. The company produces the entire range of naval combat ships, such as the nuclear-powered French aircraft carrier Charles de Gaulle, the frigates of the La Fayette class or nuclear or conventionally powered submarines.

A smaller company in naval shipbuilding is the company Constructions Mecaniques de Normandie in Cherbourg (CMN), which builds smaller combat ships, but also frigates. CMN's customers include Kuwait and Morocco.

The most important military aviation company in France is Dassault S.A., founded in 1936, which has 23,400 employees and is one of the global military aviation companies with representatives in around 70 countries worldwide. The best known products still used in the air forces of several countries are the Jaguar, Mirage 5, Alpha Jet, the Super Etendard carrier-based fighter, and the Atlantique maritime patrol aircraft. The latest developments are the Mirage 2000 fighter and the Rafale carrier-based fighter.

The main company for propulsion systems for aircraft, helicopters, satellites, rockets and missiles is Safran Aircraft Engines (formerly Snecma, Société nationale d'études et de construction de moteurs d'aviation or Snecma Moteurs) with 15,700 employees. Today, some 40 of the world's most important combat aircraft and 23 of the world's most important helicopters are powered by saffron engines.

Another military aviation company is Airbus. In several locations in France and Germany, they develop and produce fixed wing aircraft (military transport aircraft) and light to heavy transport helicopters. The most important helicopter types are the Tiger combat helicopter and the NH-90 transport helicopter.

The French defence industry has the excellent electronics company Thales, a global player with a total of 80,000 employees. It can look back on a history rich in tradition until 1892. The company is already well prepared for globalization and has a network of international cooperation with electronics companies in Europe and overseas. For example, the company works with US defence giant Raytheon, South Korean electronics giant Samsung and German and British electronics companies to solve future projects. If we look at the Thales product range, we can see 5 main areas of focus. Thales develops and produces vehicles and battlefield surveillance systems for land forces, complete shipbuilding solutions for the navy, complete solutions for airspace surveillance, innovative solutions in space technology and security technology for all possible occasions.

Another company that cannot be clearly classified in the chosen classification is MBDA France. MBDA France belongs to the MBDA Group, which produces high-quality guided weapon systems in several countries. The French part of the company is better known under its original name Mecanique Avion Traction (Matra), which has always produced guided weapons, such as the anti-aircraft guided weapon Mistral.

Despite various restructuring measures, the French defence industry presents itself as an efficient, innovative and future-oriented industrial sector capable of offering complete weapon systems for all branches of the armed forces. If the European defence industry is to become more competitive in the future, the French defence industry will play a key role in this process.

The French defence industry has always been innovative. Here in the pictures the manufacture of a submarine and the production of fighter planes

The figurehead of the French Navy is the aircraft carrier Charles de Gaulle

The figurehead of the French Air Force is the Rafale

Finnish shipyards also produce combat ships for the navy

4.8 Greece

History

The history of Greece is divided into five sections. In the pre-Greek period until the immigration of the Greeks around 1900 BC, the Hellenadic, Minoan and Cycladic cultures are of outstanding importance. With the two waves of Greek immigration, Greek history in the narrower sense begins in the following period. The third epoch begins with the assumption of power by the Macedonian Phillip II. Phillip's son Alexander conquers a world empire within 10 years. After Alexander's death, the empire disintegrated and was divided among his successors. In the next epoch Greece was conquered first by the Romans, later by the Ottomans. During this epoch, at the beginning of the 19th century, the Greek struggle for freedom took place. The Greeks gained independence in 1830. Afterwards a modern nation state was constituted. After independence there were a series of changes of throne and wars against Turkey and in 1912/13 the 1st and 2nd Balkan War, in which Greece was able to achieve territorial gains. During the First World War, Greece initially remained neutral, but later entered the war on the side of the Entente. In the period between the wars, it had to accept defeat against Turkey and to clear Asia Minor. During the Second World War, Greece came under German occupation. After the war, monarchist forces came to power. In 1967 an officer's junta took over and the last king had to leave the country.

Defence Industry

In Greece, both state-owned and privately owned companies in the defence industry develop and produce light weapons, ammunition, combat vehicles, naval vessels and modernise combat aircraft and helicopters.

The main producer of ammunition and explosives is Pyrkal, the oldest Greek defence company, founded in 1874; since 2004 Pyrkal has merged with the defence company Hellenic Arms Industries. The Hellenic Arms Industries (E.B.O.) company, which has been in existence since 1977, produces small arms, grenade launchers, guided weapon systems, ballistic protection vests in 5 production facilities with 1,300 employees and, under licence, the Artemis anti-aircraft defence system, the Carl Gustav anti-tank tube, the Milan anti-tank guided weapon and the Patriot missile defence system. Since 2004, the two companies have merged and operate under the name Hellenic Defence Systems (EAS).

Fighting vehicles are manufactured at Hellenic Defence vehicle and Hellenic vehicles. Hellenic vehicles was founded in 1972 as Steyr Hellas and was taken over by the Greek State in 1987. Since then, Thessaloniki has been producing all-terrain command vehicles and armoured combat vehicles such as the Leonidas and Kentauros infantry fighting vehicles.

The most important Greek defence company in the aviation sector is Hellenic Aerospace, which was founded in 1975 and supports the Greek Air Force in the modernization and maintenance of its air fleet in all areas from electronics to weapons systems.

Naval shipbuilding is carried out at the state-owned Hellenic Shipyard in Skaramanga on Attica, founded in 1939, and at the private shipyard Elefsis near Piraeus. While Elfsis offers landing ships and speedboats as well as comprehensive logistical services, the state-owned shipyard specializes in maintenance and new construction of guided missile speedboats and frigates.

The Greek defence industry also has the electronics company Intracom, which is engaged in the development and production of telecommunications equipment, Command&Control&Communication solutions and electronics for guided weapons.

4.9 Great Britain

History

The oldest settlement of the British Isles took place in the Paleolithic Age. Around 500 B.C. Celts immigrated and formed the warlike upper class. In the 1st century before Christ, Belgians from Northern Gaul settled in England. They offered fierce resistance to the Romans, who had crossed the Channel for the first time under Julius Caesar from 55 BC onwards. England was conquered after fierce resistance and became a Roman province. After the withdrawal of the Romans around 410 AD, administration and culture fell apart. The English called on Angeln and Saxons for help against the warring nations from the north, who subsequently became the masters. The rule of the Anglo-Saxons was replaced by the Normans around 1066. After the invasion of the Normans from France into England, both states banished a common history. The Normans organized themselves in the form of a closed feudal hierarchy. In this early period Henry II was the most powerful ruler in England. After establishing his system of rule in England and Normandy, he forced Wales, Scotland and Ireland to recognise his sovereignty. Under his successors Richard I. Lionheart and John without Land the power of the king declined. The nobility used this weakening of the king in 1215 to wring the Magna Charta from the king, which for the first time laid down in writing the rights of the estates. In the following century England fought the Hundred Years' War against France. England was driven from the mainland in this war and could only hold Calais and the Channel Islands. The weakening by the war favoured the War of Nobility, the so-called War of the Roses between the House of Lancaster (coat of arms: red rose) and the House of York (coat of arms: white rose). Henry VII of the House of Tudor restored peace and order. The rule of the Tudor was very productive for the advancement of England. Henry VIII had warships built, took an interest in foreign policy and separated from Rome and was elected head of the Anglican Church. His daughter, as Elizabeth I Queen of England, faced the conflict with Spain. The Spanish Armada was defeated and the way was cleared for England to become a naval power. The following centuries were marked by the increase in wealth due to the beginning colonisation. For a short time under Cromwell the kingship was interrupted and England became a republic for 11 years. In the wars against the Netherlands, the Netherlands was defeated and England finally rose to become a world power. In the centuries that followed, England took part in the wars on the mainland again and again, mostly leaving the scene with the victors. England probably experienced its greatest period from the middle of the 19th century onwards, when England was the non-plus-ultra of the world. English inventions in technology, physics and chemistry went around the world, England maintained the largest and most modern battle fleet and England had the largest colonial empire. The First World War was the first turning point in England's growth policy. Although England still received an increase in territory after the war, the over-expansion of the empire led to economic difficulties, which resulted in a broken economy after the Second World War. Without the help of the USA, England would probably have slipped into a long recession. The first post-war decades were also the time of the beginning of the end of the colonial empire. England lost most of its colonial empire, although the former colonial states remained united in the Commonwealth of Nations as now sovereign states. European policy and good relations with the USA became the question of survival as a world power.

Defence Industry

Military Aviation Industry

The history of military aviation in Britain begins with the deployment of a balloon force in the late 19th century. The balloons were used on a larger scale during the Boer War 1899-1902. At the beginning of the 20th century, powered flight made great strides following the successes of the Wright brothers in the USA. Flight demonstrations by the Wright brothers in Europe and flying competitions, such as flights across the English Channel and long-distance flights in England, caused interest in the new aircraft to rise sharply. As a result of the emerging rivalry with the German Reich, Great Britain, after studying the development of military aviation, decided in 1912 to establish the Royal Flying Corps (RFC), which combined all the aviation activities of the Army and Navy and had a Military Wing and a Naval Wing as well as a joint flying school in the Central Flying School. After taking over the organisation, it was decided to procure the necessary aircraft. This turned out to be a complex task, as the different needs of the army and the navy had to be taken into account. So valuable time was spent on surveys and meetings. Great Britain therefore only had around 161 aircraft and 200 trained pilots when it entered the First World War.

In mid-1914, some 15 companies in Great Britain were producing military aircraft, but even the large ones among these companies were not geared for series production. This inefficiency meant that the total annual output was only about 100 aircraft. During the first phase of the First World War, Great Britain received support from France in the form of the supply of airframes and engines. It also increased the production capacity of the airframe industry in the UK and encouraged a large number of foreign companies to build aircraft. This enabled Britain to produce a total of 55,000 aircraft during the war. This increase in production also led to an expansion of employment in the aviation industry in the broad sense. In the period between summer 1916 and autumn 1918, the number of workers quintupled from around 60,000 to around 347,000. In purely arithmetical terms, the industry's aircraft production increased from 1.5 machines per day in August 1914 to over 90 in the last days of the war in 1918.

Let us at least look at an overview of the most important military aircraft designers in Great Britain.

A.V. Roe and Company (Avro) was founded in 1909 by A.V.Roe and his brother Humphrey V. Roe in Manchester. In addition to conventional aircraft, the company also built the first British seaplanes. In 1913 the company was renamed A.V. Roe& Company Ltd. and was given the trade name Avro. Avro's most famous aircraft during the First World War was the Avro 504 training aircraft. 8,340 of this type were produced. Avro survived the difficult times after the war and worked on the development of long range aircraft and autogyros. Before the Second World War the company built bombers. One of the probably best bombers of the war, the four-engined long-range bomber Lancaster, was a creation of the Avro company. In 1945 Hawker Siddeley took over the military aviation division of Avro companies. The last military aircraft from the Avro design office was the Vulcan bomber. With the reorganization of Hawker Siddeley, the trade name Avro also disappeared in 1963.

Airco was founded in 1912 by George Holt Thomas in Hendon as an aircraft manufacturing company and initially produced French Farman aircraft under license. In 1914, Geoffrey de Havilland was hired as head of development and since then the company has built its own designs. The most important models during the First World War were Airco D.H. 2 (401 pieces), D.H. 6 and D.H. 9 (2,280 pieces). After the war the company got into economic difficulties and had to go bankrupt in 1920. The production facilities were bought by Birmingham Small Arms. With the rest of the equipment DeHavilland founded the DeHavilland Aircraft Company.

Handley Page was founded by Fredrick Handley Page in 1909 as Handley Page Aircraft Company. In 1915 the largest British aircraft of the time, the type O/100, was built. This type was improved and afterwards used as O/400 between 1918 and 1919 as a long-range bomber. A total of 450 of them were produced. In the interwar period the company mainly produced civil aircraft. A number of bombers were developed for the military, such as the Halifax, Heyford and Hyderabad types. After the Second World War, Handley built the Victor bomber. During the development of the Jetstream, the company overstretched itself financially and had to go bankrupt in 1970. However, the Jetstream continued to be built by British Aerospace.

The Sopwith Aviation Company was founded in 1912 by Sir Thomas Octave Murdock Sopwith and Fred Sigrist. Sopwith constructed one of the most successful aircraft of the First World War, the Sopwith Camel. A total of 5,734 of them were produced. Further models of importance were the Sopwith Dolphin (1,532 pieces), Sopwith ½ Strutter (5,720 pieces), Sopwith Snipe (497 pieces) and Sopwith Pup (1,707 pieces). After the war the company went bankrupt. Soon after, Sopwith and his chief designer Harry Hawker founded H.G.Hawker Engineering (see under Hawker).

Vickers produced fighter planes, bombers and also passenger planes between 1911 and 1962. The most important aircraft type of the First World War from the Vickers company was the F.B.5 lattice-tail fighter, which was built in a number of 224 pieces. After the war, Vickers also built bombers for the Royal Air Force, including around 11,400 medium-weight Wellington bombers, which were used in the Second World War.

The Bristol Aircraft Company was founded in 1910 by George Stanley White near Bristol. During the First World War a total of 3,101 F2B ground support aircraft were built. In the interwar period, Bristol developed the night fighter (around 6,000 Beaufighters), the torpedo bomber Beaufort and the medium-weight bomber Blenheim. All these models were used with great success in the Second World War. After the war, Bristol was split into two companies, Bristol Aircraft and Bristol Aero-Engines. Bristol Aircraft merged with English Electric and Vickers in 1960 to form the British Aircraft Corporation (BAC). The aircraft engine company Bristol first merged with Armstrong-Siddeley in 1959 to form Bristol Siddeley Engines Ltd. and in 1961 acquired DeHavilland Engines and Blackburn Engines before being taken over by Rolls Royce in 1966.

The Royal Aircraft Factory was founded in 1892 in Aldershot and Farnborough. Aircraft design and production began in 1911, and it should be noted that the Royal Aircraft Factory was mainly a design office rather than an efficient production facility. Many of the Royal Aircraft Factory's designs were produced by other aircraft manufacturers. Among the most important aircraft designs were the E.E. 2 (Bleriot Experimental)/ 3,535 units built, the S.E. 5 (Scout Experimental)/ 5,205 units built, F.E. 2b (Farman Experimental)/ 2,190 units built. From 1918, the Royal Aircraft Factory ceased production and turned its attention to research.

Supermarine was founded in 1913 by Noel Pemberton-Billing as a sea plane factory. During the First World War Supermarine also built a four-wing aircraft, which was intended for defence against German zeppelins. The Supermarine P.B.29 had a recoilless cannon as armament and the Supermarine Nighthawk had an additional power source for a powerful searchlight. In 1938 Supermarine was taken over by Vickers Armstrong. Even the first construction of a land-based aircraft became one of the greatest successes in military aviation history, the Supermarine Spitfire, the backbone of British fighter aviation during the Second World War. Around 20,000 copies of the Spitfire were produced. When Vickers was transferred to the British Aircraft Corporation, the trade name Supermarine also disappeared from the register of aircraft manufacturers.

DeHavilland was founded in 1920 by Geoffrey DeHavilland. DeHavilland continued in his own company the work he had successfully started at Airco (see there). With his own engines he succeeded in the construction of some record aircraft, such as the Gipsy and the Tiger Moths. DeHavilland began experimenting with wood as a material for airframes at a very early stage. The best construction based on wood was the Mosquito, a bomber, fighter and low attack plane. About 7,780 of the Mosquito were built. After the Second World War, DeHavilland developed jet-powered fighter planes for the army and navy. Probably the most famous examples were the DH 100 Vampire and Sea Vampire, DH 110 Sea Vixen and the DH 112 Venom and

Sea Venom. DeHavilland was bought by Hawker Sidddeley in 1959 and has been part of British Aerospace since 1977. De Havilland also established companies in Canada and Australia. De Havilland Australia was taken over by Boeing and is called Boeing Australia. DeHavilland Canada was taken over by Bombardier. In Canada the passenger aircraft Dash-7 and Dash-8 are built.

Short goes back to the foundation of an aircraft factory by the Short brothers, who built aircraft of the Wright type under licence from 1908. From 1911 onwards, Short produced its own developments. The best known in-house development is the Short 184 bomber, which was used during the First World War. In the interwar period Short dedicated itself to the development of bombers and flying boats. The most famous creations are the Bomber Short Stirling (2,375 units built) and the Short Sunderland flying boat (700 aircraft built). After the war, production continued mainly at the Belfast plant with the production of transport aircraft (Short Skyvan). Bombardier took over the company in 1989. Since 1993 the Belfast plant was renamed Short Missiles (50% Bombardier, 50% Thomson-CSF). Since 2001, the company is solely owned by Thales.

The aircraft manufacturer Gloster was founded in 1917 by Airco as Gloustershire Aircraft in Cheltenham. During the First World War, the company produced mainly Airco, Nieuport and Bristol aircraft. The output was around 45 aircraft per week. In 1926, as the name was almost unpronounceable to foreigners, the company was changed to Gloster Aircraft Co. In 1934 Gloster was taken over by Hawker. In the interwar period, well-known types were developed, such as the Gloster Gladiator. It was the last fighter of the Royal Air Force in double-decker design. It also proved its worth as a version for use with aircraft carriers. England's first and only jet fighter of the Allies, which was used before the end of the war, was the

Gloster Meteor, which was used in the last days of the war, although its development had already begun when the war began in Europe.

Saunders-Roe, abbreviated Saro, is an aircraft company based on the Isle of Wight. The name dates back to 1929, when Alliot Verdon Roe and John Lord took control of the ship and aircraft building company S.E. Saunders. The company specialized in the construction of flying boats. After the era of flying boats came to an end, the company first attempted to produce helicopters and later turned to the development and production of hovercraft from 1959. In 1959 Saro was taken over by Westland Aircraft. In 1964 the Saro hovercraft was merged with Vickers Supermarine Hovercraft to form the British Hovercraft Corporation.

Armstrong Whitworth was founded in 1913 by William Armstrong and Joseph Whitworth as the aerospace division of the armaments company founded in 1897. During the First World War, the aircraft factory produced a series of bombers. The best known was the model F.K. 8 (1,652 pieces). The aircraft factory merged with Siddeley to form Armstrong-Siddeley in the mid 1920s. In the course of the concentration of the British aviation industry, Armstrong-Siddeley was merged with Hawker to form Hawker Siddeley.

Sopwith founded H.G.Hawker Engineering with his chief designer Harry Hawker in 1920 after the bankruptcy of Sopwith. During the Second World War, Hawker aircraft, together with Supermarine Spitfires, were the backbone of the British Fighter Air Force. The most important types were Hawker Hurricane (around 14,500 built), Typhoon and Tempest (3,300 built). After the Second World War, the company continued to assert itself on the market. Hawker, now part of the Hawker Siddeley Group. The Hawker Siddeley Group was thus the third major aviation group in Great

Britain, alongside the British Aircraft Corporation (BAC) and Westland. The Hawker Siddeley Group was a very successful group; it developed the Buccaneer, the Nimrod, the Hunter and the Harrier vertical take-off aircraft, among others. In 1977 the Hawker Siddeley Group merged with British Aircraft Corporation (BAC) and Scottish Aviation to form British Aerospace (BAe).

The Spanish Don Juan de la Cierva was a pioneer in the development of helicopters. He emigrated from Spain to England in 1925 and in 1916 founded his own company dedicated to the development and construction of helicopters. In England he developed the Cierva Autogiro C-6 on the basis of the Avro 504. In 1928 he succeeded in crossing the English Channel with the development of the C-8. The C-8 was already being produced in large numbers. Cierva's last big development was the C 30A, an already well maneuverable autogiro.

The UK currently employs 140,000 people in its defence industry and a further 110,000 in its suppliers. The flagship of the UK defence industry is **BAE Systems**, an international company engaged in the design, manufacture, marketing and support of complex weapons and weapon systems for land, sea and air forces. The company develops and manufactures battle tanks, combat ships, combat aircraft, radar systems, communications equipment, electronics and satellite technology and has subsidiaries on all 5 continents with a total of more than 85,800 employees. So far, the company has sold its products in about 130 countries. Since 2018 the company has been divided into Electronic Systems, Cyber & Intelligence, Platforms & Services (US), Air and Maritime.

In the area of land systems, the two most important subsidiaries are Alvis Vehicles and RO Defence. RO Defence develops and produces small arms, artillery systems, for example a 155mm howitzer with high tactical manoeuvrability, and marine guns and the corresponding ammunition types. It also produces infantry combat vehicles, such as the Warrior, a tracked vehicle with a 40mm cannon and high tactical mobility. Alvis Vehicles is one of the leading suppliers of all types of combat vehicles, ranging from main battle tanks and armoured combat vehicles to light infantry fighting vehicles. Alvis is also the owner of the Swedish military vehicle manufacturer Hägglund, where the battle tanks are also manufactured.

The Navy has always been of particular importance to Britain, with a centuries-old naval tradition ranging from the successful fight against the Spanish Armada and Nelson's victory at Trafalgar to the Falklands and recent deployments in Iraq. In the field of naval systems, BAE develops and produces a wide range of weapon systems from combat ships and submarines to naval ordnance. Particularly noteworthy here is the planning work for the new aircraft carrier class, landing ships and submarines. BAE is also the prime contractor for the production of the Type 45 destroyer.

BAE also develops and produces ultra-modern weapon systems for the German Air Force. For example, it was involved in the development and production of the Tornado multi-purpose combat aircraft and is involved in the Eurofighter project and the development of the Joint Strike Fighter. BAE is also a successful supplier of the Harrier vertical take-off and landing aircraft and the Hawk training aircraft.

BAE is also successfully active in the fields of communications technology and space travel. It develops network-oriented solutions for C4ISR (Command&Control, Communications, Computing, Intelligence, Surveillance and Reconnaissance) and is integrated in the European and also US space program. BAE is also prepared for future challenges and operates a center for operational research and defense analysis.

Rolls-Royce serves the civil aviation, military aviation, marine technology and energy market segments worldwide. Today, a total of around 54,000 Rolls-Royce gas turbines are in operation worldwide. With more than 500 airlines, 4,000 operators of business and commercial aircraft and helicopters, 160 armed forces and over 2,000 naval customers including 50 navies, Rolls-Royce has a large customer base. The company employs 50,000 people worldwide. In the defence sector Rolls-Royce is involved in the most important fighter aircraft developments Eurofighter and Joint Strike Fighter. As the world's most important engine manufacturer, Rolls-Royce supplies engines for Lockheeed Martin's new C 130J transport aircraft and also plays a major role in the development of the European military transport aircraft A 400M. As market leader in the marine segment, Rolls-Royce manufactures the engines for the most important new military programs in the U.S. and U.K. and also offers new products to civilian customers.

Another successful company in the defence industry is the **Smith Group,** which is regarded worldwide as one of the most important suppliers of security technology, medical technology, aviation technology and special technologies, such as complex naval systems.

An important company in the shipbuilding sector is the **Babcock Group**, which acquired the VT Group (formerly Vosper Thornycroft plc). Vosper Thorncroft was formed from the merger of Vosper and Thornycraft, which have their roots in the 19th century. The company specializes in the development and manufacture of destroyers, speedboats and the provision of logistics and communications technology and training support.

Another important company with a strong defence industry is **GKN**, which has 58,000 employees in some 30 countries and generates revenues through the development and production of Westland civil and military helicopters and through participation in civil and military aircraft programmes such as the Airbus programme or the F-22 Raptor, C-17 Globemaster or F-35 Joint Strike Fighter.

A smaller company in the aviation industry should not go unmentioned. The **Britten-Normann** Group, owned by the Zawawi family from the Sultanate of Oman, produces highly mobile aircraft, for example the Islander and Defence types, which are used by both civil and military customers.

A major British company with a long tradition in the production of weapon systems is **Qinetiq** with its 6,000 employees. Oinetiq was part of DERA, the Defence Evaluation and Research Agency of the Ministry of Defence, until 2001. After the demerger, an organization DSTL remained with the Ministry of Defense to handle sensitive research contracts, and Oinetiq became a spin-off company that can offer tailor-made solutions for land systems, naval and aerospace applications and network-centric warfare.

Insys, which was acquired by Lockheed Martin UK and which is one of the most important providers of innovative solutions in communications technology, in particular solutions to problems related to the use of C4ISTAR (Command&Control, Communications, Computing, Intelligence, Surveillance and Reconnaissance) at tactical, operational and also strategic level, was also specialised in innovative solutions.

Over the last decades **Cobham** has become a leading provider of military aviation-specific services. The company can provide solutions to communication, training and education and avionics related issues.

The list could be continued for several more pages, but the author omits this for reasons of space and asks the unnamed companies for their understanding. However, it should also be mentioned that the British defence industry is strongly networked with the US defence industry and the most important US companies in this sector have branches in Great Britain.

All in all, the British defence industry presents itself as an efficient and innovative branch of industry, which always strives to offer the latest technologies through close cooperation with both European and US partners. In the context of a merger of all European defence industries to form an efficient defence equipment manufacturing industry with worldwide sales opportunities, the British defence industry will play a key role in the transatlantic partnership through its special bridging function.

At the height of power, British armament factories produced all weapons for the army, air force and navy

The Warrior was a highly successful armoured infantry fighting vehicle from the British defence industry

UK shipyards are able to build all combat ships

The defence industry in Greece is diverse. And also produces under licence military equipment developed by foreign companies

4.10 Italy

History

According to tradition, Rome was founded in 753 BC. Already in the 3rd century before Christ Rome ruled almost all of Italy. The Roman Empire experienced its greatest expansion in the 2nd century after Christ. After the end of the Roman Empire in 476 A.D., Germanic empires were founded in the territory of present-day Italy. The last rulers in a chain of Germanic rulers were the Staufer. After the execution of the last Staufer Konradin in Naples (1268), the Germanic rulers were replaced by French rulers. An Italian system of states only formed in the 14th and 15th centuries. The republics of Venice, Genoa, Florence as well as the principalities of the Visconti and Sforza in Milan, the Este in Modena, Scaliger in Verona etc. were established. In the middle of the 15th century there were five central states (Naples-Sicily, Florence, Papal States, Milan, Venice), under whose protection the smaller states could also exist. At the end of the 15th century there were power struggles between the European great powers France, Spain, German Habsburgs for supremacy in Italy. The power struggles led to a division of the peninsula Italy, only Sardines and Genoa remained as the only state under a national dynasty. From 1830 onwards the Risorgimento and the unification of the country took place. At the latest since C. Cavour had taken foreign policy into his own hands, the unification process was accelerated. In 1859 Lombardy was won, in 1866 Veneto and finally in 1870 the Papal States. After the unification, Tripoli and the Cyrenaica were acquired. After the First World War, the Brenner border, Gorizia, Trieste, Istria and parts of the coast and Zara were won. By marching on Rome in 1922 Mussolini forced the king to give him the government. Sanctions were imposed by the League of Nations for the invasion of Ethiopia. Italy therefore approached the German Reich and in 1936 entered into an alliance with the German Reich and Japan. At the beginning of the Second World War, Italy initially remained neutral, but then took part in the war on the

German side. After an unsuccessful war effort, Italy concluded an armistice in 1943 and fought against the German Reich. Through the Peace of Paris, Italy lost its colonies and also the territories in Istria and the coastal lands.

Defence Industry

At the time of the agreement, the national income was not quite a third of that of the French and only a quarter of that of the British. Industry accounted for a very small proportion of total national income: no more than 20.3% of GDP, compared with 57.8% for agriculture and 21.9% for trade, transport and public services. The fact that the South was not favoured by the beginning of statehood is shown by the wave of industrialisation in the North in the 1980s and 1990s. In the North, there was a considerable shift to heavy industry - iron, steel, shipbuilding, and later the automobile and textile industries. Italian industry grew rapidly, and agriculture in the North showed similar success, while in the South agriculture remained dominant. But as soon as the Italian statistics are put into a comparative perspective, the glamour quickly fades. An iron and steel industry did develop, but in 1913 its production level was only about one-eighth of the British level and even two-fifths of the Belgian level. Growth in Italy was impressive, but it started from a low initial level, so the real results were not impressive. In addition, governments failed to compensate for austerity measures and tax increases, which led to a sharp deterioration in the living conditions of the poorer classes, further increasing the plight of the workers. For example, at the beginning of the First World War, it had not even reached a quarter of the industrial capacity of Great Britain and its share of world production was only 2.5 percent. Although Italy was one of the great powers, it is remarkable that any other power had industrial capacity two or three times as large. Some were six times (German Empire and Great Britain) and the USA even thirteen times stronger. Moreover, there was a general slowdown in growth even after 1908. Demand for capital goods declined, while the metallurgical, chemical and mechanical engineering industries countered a crisis of overproduction. Since a large part of the investments were made through loans and these could no longer be repaid by the companies, the Italian central bank had to intervene several times and service bad loans. The consequences for Italy's strategic and diplomatic position were therefore depressing. The Italian general staff was not only aware of the numerical and technical inferiority of its armed forces against France and Austria-Hungary, but also realized that the industrial conditions and the railway network were insufficient for warfare.

Due to the structure of the economy, the outbreak of the First World War led to state intervention in the entire economy, which was unprecedented in the history of the united Italy. The state guaranteed and directed total mobilisation, disciplined production and suppressed strikes, encouraged concentration processes and paid excessive prices in order to obtain the necessary cannons, guns, trucks, planes, uniforms and provisions in the shortest possible time. Under these conditions Ansaldo doubled its workforce from 50,000 to 100,000 employees between 1914 and 1918 and increased its capital tenfold between 1916 and 1917. Fiat and Breda experienced similar developments. In spite of all the efforts made, Italian industry was only able to maintain the level of industrial and agricultural production and thus to supply the troops and the population to a certain extent, thanks to the generous aid of the Entente powers. This is confirmed by the trade statistics of the war years: In 1914, Italy imported for 2.9 billion lire, in 1915 for 4.7 billion, in 1916 already for 8.4 billion, in 1917 for 14 billion and finally in 1918 for 16 billion, while exports remained at a level close to 3 billion. At the end of the war, this annual negative trade balance led to a mountain of debt of around 3 billion dollars, mainly to the USA and Great Britain. These enormous debts placed a particularly heavy burden on the Italian economy after the war.

After the war, the difficulties in the transition from a war economy to a peace economy, the fall in productivity and the crisis that hit first Ansaldo and then the other large trusts also affected the solvency of the credit institutions. The State was forced to intervene by way of a steering mechanism and, by means of the Banking Law of 1936, to separate the activities of the banks that financed specifically for industry from those that handled the normal lending business. The former were specially protected by the state.

In the interwar period Italy always had a negative trade balance, which led to a shortage of foreign exchange, making it impossible to import high quality machine tools and also to create raw material warehouses with a long reach, so that only warehouses with a reach of 3 to 4 months were available. In addition, there were annual budget deficits because up to 30 percent was spent on the military. This was compounded by the dependence on raw materials, which were only available to a limited extent after the beginning of the Second World War. This led to a decline in the production of basic materials for industry, for example, steel production fell by around 25 percent in 1943. The result was a dramatic decline in the most important branches of production; for example, the shipping industry was only able to reach 15 percent of its pre-war level. Aircraft, tank and vehicle production was also affected, however, with the result that the Italian army was only able to rely on some   aircraft   tanks. Due to the lack of precession machines the Italian industry was not able to produce armaments of similar quality as the German Reich and also the quantity lagged far behind the planned figures, so that the equipment of the armed forces remained rather modest compared to the great powers of that time. All the more remarkable is the fact that Mussolini went to war with this poorly equipped armed forces. Without battle-ready tanks, anti-aircraft guns, aircraft carriers, radar and without an adequate supply system, Mussolini plunged his country into a war of the great powers in 1940, assuming he had already won. The whole emphasis on arms and numbers obviously ignores the factors of leadership, the quality of the crews and the national will to fight, but the fact was that these elements did not remotely compensate for Italy's material shortcomings, but instead contributed to its relative weaknesses. Despite the superficial fascist indoctrination, nothing had changed in Italian society that would have made the army an attractive career for talented men.

The production of agricultural products was in a similar situation to heavy industry. Production fell continuously during the war to a level of around 25 to 30% of the pre-war level. One of the main reasons for this was the lack of manpower, as Italy greatly increased its army during the war (up to 8 million soldiers) and could only rely on a few prisoners of war. Moreover, Italy could not access foreign resources, as was the case with the German Reich, which occupied large parts of Europe. Moreover, Italian agriculture was less mechanised, so that labour could not be replaced by the use of machines. It was not possible to advance mechanisation because tractor production had to be put on hold in favour of motor vehicle production. The result was rationing of food, which only provided 1000 to 1100 calories for the adult. After the armistice the supply of the population became even more drastic until the end of the war, because the German Wehrmacht acted as an additional demand for food.

Italy had armoured cars in the First World War. The I.Z. was manufactured by Ansaldo on a Lancia chassis. In 1915 20 of them were produced. The vehicles were armed with three 6.5 mm machine guns, which were housed in two towers. The vehicles were still in use in East Africa and Libya at the beginning of the Second World War.

Italy initially did not build its own tanks during the First World War, but imported the French type Renault FT. Italy started to build a tank based on this concept. Before the end of the war Fiat developed the Tipo 2000 Modello (combat weight: 40t), of which only 2 prototypes were available in 1918. Of the much lighter successor model Tipo 3000 (combat weight: 5.5t), 100 units were produced, which joined the troops in 1923. Further light tanks came from the CV (Carro Veloce) 33 and 35 family, of which a total of about 2,000 were built. The Italian CV 33 and 35 armoured vehicles were used as accompanying tanks for the infantry. In this capacity they were quite useful, but when they were used as battle tanks - against all common sense - they failed miserably. When Italy entered the Second World War, it had about 700 tanks. Since the Italian philosophy of tank construction until 1940 was limited to the production of light vehicles under 14t, the Carro Armato M.11/39 (combat weight: 12t, about 1,000 of them were built in total) and the M.13/40 (combat weight: 14t, about 800 of them were built in total) and the Type M.14/41 (combat weight: 14.5t, about 1,100 of them were built), formed the backbone of the tank weapon in World War II. Since both proved to be inferior to the Allied tanks, it was decided that a heavy tank would have to be built. So a vehicle of 25t with 14mm front armour and a 7.5cm cannon was ordered in 1940. However, the resulting tank had a long and arduous history, so that in 1943 only 21 pieces of the type P.26/40 were available. In total Italy produced only 2,000 tanks during the war; little compared to the German Reich, which produced about 66,000 tanks of all types.

A special type of armoured vehicles were the Semovente (assault guns). Between 1940 and 1945 a total of 1240 Semovente of different configurations were produced. The Semovente had a chassis based on the chassis of the medium battle tanks and main armament of different calibres. The main types were the Semovente 75/18 with a 7.5 cm gun and the Semovente 105/25 with a 10.5 cm gun. The climax of the development of the Semovente was the Semovente 149/40 with a 14.9 cm gun. However, only one prototype had been completed by the end of the war.

Italy was the first country in the field of military aviation: as early as 1884, an army flying unit equipped with balloons was set up there. When Italy entered the First World War against Austria-Hungary in May 1915, its air force was probably better trained than that of the enemy. However, this was compensated for to a certain extent, as mostly used French types such as Nieuport, Bleriot or Maurice-Farman aircraft were used. During the war, they also used their own developments, such as the Caproni. During the First World War, aircraft were built in seven factories, so that most of the 1,778 aircraft available at the end of the war were of domestic production.

In the interwar period Italy produced aircraft that set world records, but when it entered the war in 1940 - at that time Italy had 3,269 aircraft of various types, of which only 1,795 were combat-capable - the squadrons of the Regia Aeronautica still included a large number of biplanes and high-deckers from Fiat, Breda and Caproni. During the war a number of new aircraft were produced, such as Macchi aircraft (1100 Macchi C 202 Folgore single fighter, 310 Macchi C 205 Veltro), Reggiane (250 Reggiane 2001 Falco single fighter, 250 Reggiane 2002 and 36 Reggiane 2005 Sagittorio), Fiat (785 Fiat G.50 Freccio), Cant (380 pieces Cant Z.506 Airone, Cant Z 1007 to Alcione, Cant Z 501 Gabbiano),Savoia-Marchetti (1370 pieces medium range bomber S.M.79 Sparviero, 535 pieces S.M.81 Pipistrello and 411 pieces S.M.82 Canguro) and Piaggio Piaggio produced in Pontedera 163 pieces of a four-engined bomber type Piaggo P.108 B with long range, which could transport a bomb load of 3500kg over 4000km, but they were only used in small numbers and too late. Altogether Italy produced about 11,500 aircraft during the war; little compared to the German Reich, which produced about 117,000 aircraft.

The "Regia Marina" has always played an important role for Italy, as only it could secure the sea routes and thus guarantee the sea trade which is vital for the economy. Around 1860, the Italian navy had only one liner, 10 screw frigates, 5 screw corvettes and smaller units. Already after becoming a state, the old naval tradition was remembered and armoured ships were purchased from France, the USA and England, but thanks to talented engineers such as Benedetto Brin, efficient naval shipyards were built in La Spezia (Vickers-Terni), Orlando and Odero in Genoa. The guns for the ships were manufactured in the Armstrong factory in Pozzuoli. At the beginning of the First World War, the ambitious fleet construction programme resulted in the sixth strongest fleet in the world, far stronger than the Austro-Hungarian navy. The interwar period brought a slump in shipbuilding due to the fleet limitation treaties, but Mussolini supported the navy because he saw in it an important means of power. The allocation of resources to naval shipyards was therefore given special importance. Battleships and cruisers were built, but the construction of aircraft carriers was completely neglected because Mussolini had allegedly banned it. Only one aircraft carrier was developed, but it was no longer used, which was to have a very fatal effect on the Italian battle fleet during the battles for Malta and Cape Matapan. The Italian fleet (about 500,000 BRT and thus in 5th place in the world) with its 6 battleships, 20 cruisers, 53 destroyers, 68 torpedo boats and 111 submarines represented a considerable power factor in the Mediterranean, but their use was, as far as the battleships were concerned, only possible to a limited extent without carrier-supported hunting protection on the high seas. Moreover, the surface combat ships proved to be insufficiently protected and only the battleships of the Vittorio Veneto class had radar equipment. The submarines also showed serious structural deficiencies, so that Italy lost a large part of its submarines during the war.

The armament and equipment of the Italian infantry of the late 19th century was similar to that of the rest of Europe. At first, rifles and machine guns were produced under licence, but very soon they were being developed by

Carcano, Beretta, Breda and Brizia. Although the Italian guns did not achieve anything outstanding, the Italians had one of the best submachine guns of all times - the Beretta. Little known is that the Italians were already using submachine guns in their infantry units in 1918. This weapon was constantly improved. As model Beretta 1938A caliber 9mm it was used from 1938 on in larger numbers also in World War II by the German Wehrmacht and Romania. The production of the gun was stopped only in 1950.

For a long time, Italian industry was not able to supply the army with sufficient light, medium and heavy guns and so the Italian army was almost completely dependent on foreign gun material. Only the companies OTO and Vickers-Terni were only concerned with the licensed production of foreign constructions and the Turin Arsenal developed a mountain gun. An additional disadvantage was the introduction of a barrel recoil gun by the Italian Army at the beginning of the 20th century, as all countries were already using barrel recoil guns at that time. Production was soon discontinued and foreign products were bought in. At the beginning of the entry into the First World War, only 112 guns of 149mm and 210mm calibre were available. During the First World War, French guns and the company's own mountain guns were used. After the First World War, large quantities of captured guns, some of which were later converted, were used by the artillery units. Only in the second half of the thirties new guns of own production joined the troops. However, these were by no means sufficient to equip all units with new material, so that the Italian army entered the Second World War with 7,970 artillery pieces. However, only 246 of these guns had been built after 1930.

The birth of the automobile in Italy dates back to 1890, when the Milanese entrepreneur Giuseppe Ricordi introduced the first automobile with a combustion engine to Italy. In 1899 the Fabbrica Italiana Automobili Torino

(F.I.A.T.) was founded, which soon became the most important producer. Lancia followed in 1906 and Alfa Romeo in 1910. With the threat of the First World War and the simultaneous development of civilian automobile transport, the interest of military commanders in the motor vehicle became ever greater. Large automobile manufacturers, such as FIAT, were only too willing to meet the requirements of the armed forces. C.F. Zampini-Salazar recalls in his book "Ottant Anni di Camionfiat" (80 years of FIAT trucks) that during a military manoeuvre in 1911 on the Po between Alessandria and Santhia, two complete motor vehicle compartments were set up for the first time, one for each enemy. These departments consisted of 83 cars, 97 trucks and 99 motorcycles. When Italy declared war on Turkey on 29 September 1911, FIAT was commissioned to supply light trucks with a payload of 800-1000 kg. Within a short time FIAT delivered 200 trucks 15 Bis with the designation Libya. The First World War also presented new challenges to the car manufacturers, and they produced staff vehicles, special vehicles for the anti-aircraft troops, artillery, supply troops, medical troops. Among other things, the army had 1600 FIAT vehicles with the troops.

In the inter-war period, the automotive industry had already been prepared for wartime production since 1932. During the war, the car manufacturers FIAT, Alfa-Romeo, Bianchi, Isotta- Fraschini, Lancia and OM built a wide range of military vehicles, including a standardized truck. The most important staff vehicles were the FIAT Tipo 508 Balilla and Tipo 518 Ardita in the early 1930s. Between 1937 and 1939 the FIAT 508 C Col. was built and between 1939 and 1945 the FIAT 508 C Mil. 1100. In 1938 about 620 pieces of the FIAT 2800 Mil. Col were produced. Alfa Romeo, Bianchi and Lancia also supplied staff vehicles to the armed forces. Alfa 152 pieces of the 6 C 2500 Col., Bianchi the types S6M and S9M and Lancia the Aprilia. Alfa Romeo suffered particularly badly from the effects of the war, as the plants were subjected to three bomb attacks; for example, a bombing raid on the Portello plant in 1944 led to a production stoppage until the end of the war. FIAT delivered the 508 C to light trucks. Mil., which was based on the passenger car, OM (Officine Meccaniche SpA of Brescia) Autocaretta and FIAT SPACL 39, a standardised truck of the Autocarro Unificato Program was produced by Alfa Romeo, FIAT, Isotta-Fraschini, Lancia and OM. The most famous military truck was probably the Lancia 3Ro. During the Second World War the 3Ro was used in Russia and Africa. The body of the 3Ro was extended with a 9 cm gun due to the area of use. However, the production of military vehicles had a negative effect on the production of civilian vehicles, so that there was a steady decline in the production of civilian vehicles from 56,000 in 1939 to 9,000 in 1942.

The Italian automotive industry also produced half-track vehicles and tractors for artillery. FIAT and La Moto-Aratrice tractors were already being used by the Italian artillery in 1909. During the Second World War half-track vehicles and tractors from FIAT and Breda were used.

In addition to automobiles, trucks, tractors and half-track vehicles, the Italian automotive industry also produced around 1200 armored scout vehicles, so-called Autoblinda. Several types were produced, including 24 units of AB 40, 650 units of AB 41 and 80 units of AB 43, Fiat 611 and the Lancia Lince. About 250 of the Lince, which looked very similar to the British Daimler Dingo, were produced. The vehicle had a 2.5 cm thick armouring. Remarkable was the all-wheel steering, which gave the Lince a remarkable agility. The armoured scout cars had different armament, mostly a 2 cm cannon and an 8 mm machine gun. A particularly successful construction was the Sahariana, an agile Jeep-like vehicle.

The invention of the telephone and radiotelegraphy in the 19th century also aroused great interest among military planners and led to drastic innovations in troop command and control. In all modern armies of Europe these new command and control devices were introduced almost simultaneously and the national industry was encouraged to produce these devices itself. In Italy the invention of radiotelegraphy is closely linked to the name of Guglielmo Marconi.

During the First World War, the Italian army was therefore equipped with field telephones of its own production and radiotelegraphy was also used on land and at sea. During the interwar period and also during the Second World War, however, German equipment had to be used during the campaign in Abyssinia due to production bottlenecks.

In addition, self-developed radars were used on the battleships of the Vittoro Veneto class, other ships did not have any radars. This had a particularly disadvantageous effect, since the opposing British naval forces had radars on almost all combat ships.

The overview essay on Italy's war and armaments economy from 1870 to 1945 clearly shows the major weaknesses of the Italian economy, both in terms of catching up with the rest of Europe in terms of economic development and in satisfying the needs of the civilian population and the armed forces in equal measure. The result was a production output that could not compare, either quantitatively or qualitatively, with the performance of the war and armaments economy of other major European powers. Compared to the great powers of the late 19th and early 20th century, Italy's economic performance remained rather modest, which was also reflected in the equipment and furnishings of the armed forces. Italy therefore entered two world wars without any preparations and the country did not succeed in establishing an independent war economy. Thus, Italy's entry into the First and also the Second World War was more of a burden than an advantage for the allies. The consequence was that the allies had to support not only the armed forces but also the home front with aid deliveries, which was easier to do in the First World War due to the almost unlimited resources of the USA than in the Second World War due to a German Reich, which itself had already been experiencing great shortages of raw materials and food since 1942.

In the mid-1980s, the Italian defence industry employed around 85,000 people in 500 companies - currently there are only 50,000 workers in around 90 companies and they produce weapons and weapons systems for all three branches of the armed forces. In the following, the most important companies will be presented.

The Italian defence industry for the production of land and air systems and electronics is Leonardo. The company unites some important producers. The Aermacchi company, with its main plant in Venegono, is considered one of the world's best manufacturers of turboprop and jet engine training aircraft. To date, some 2,000 aircraft have been exported to 40 countries over the last 40 years. In addition to the production of military training and light earthfighter aircraft, the company also produces in the civil segment as a supplier for the European Airbus program.

Another company in the aviation industry is the world-renowned helicopter manufacturer Augusta. Founded in 1907, the company has been producing helicopters since the early 50s. In addition to its own developments, for example the A-129 Mangusta, it also builds under licence and is involved in multinational developments, such as the European helicopter NH 90.

Participation in the Eurofighter project is of particular importance for the Italian aviation industry. The long-established aviation company Alenia Aeronautica, a subsidiary of Leonardo, which was formed by the merger of Aeritalia and Selenia, has a 21% stake in the Eurofighter project. Final assembly of the Italian 121 Eurofighters was carried out at the factory in Caselle near Turin. Besides the Eurofighter, Alenia is also involved in other multinational aircraft projects, such as the F-35JSF (Joint Strike Fighter) project. It also produces the C-27 J Spartan transport aircraft and the ATR 42 MP observation and surveillance aircraft.

The development and manufacture of battle tanks has a long tradition in Italy, although it was not possible to produce competitive products during the Second World War. After the war, the production of light and medium armoured infantry fighting vehicles became the main focus. Furthermore, the German Leopard 1 main battle tank was produced under license. The new Ariete main battle tank was jointly developed by Iveco Fiat and Oto Melara. The Italian Army has ordered 200 of them. Together with Oto Breda, Iveco developed light and medium infantry combat vehicles, such as the 8x8 Centauro family of combat vehicles. Iveco is also involved in the development and production of a family of all-terrain small and medium-sized trucks.

Italy has a long tradition in shipbuilding. Until the end of the Second World War, aircraft carriers, battleships, heavy cruisers, destroyers and submarines were built at Italian shipyards, albeit often of poor quality. Thus the Regia Marina became the most important part of the armed forces and could at least guarantee regional maritime supremacy. The most important producer of combat ships and submarines today is Fincantieri, a company that was already established in 1959 as Societa Finanziaria Cantieri Navali-Fincantieri S.p.A. and since 1984 has been a group of 8 companies involved in shipbuilding. Fincantieri has completed the latest Italian aircraft carrier, the Cavour, and plans to develop a new generation of amphibious landing ships.

Another company that produces naval ordnance is Whitehead Alenia Unterwassersysteme, a subsidiary of the Leonardo Group, which can look back on a long tradition - after all, Whitehead founded its own company in 1875 in what was then known as the Imperial and Royal Navy. In 1875 Whitehead founded the first torpedo factory in Fiume, which at that time belonged to the Imperial and Royal Monarchy, and specialized in the development and production of underwater systems. The production range includes light and heavy torpedoes, torpedo defense weapons, sonar equipment and mine countermeasures.

MBDA, a joint venture between Airbus, Leonardo and BAE Systems, develops and produces air-to-air, air-to-ground, ground-to-air missiles, anti-ship missiles and simulators. In the development of weapon systems, the company combines the entire technical know-how of the European missile industry and thus ranks second in the world as a supplier of such weapons directly behind the US high-tech company Raytheon. MBDA currently has 32 missile programmes to offer and 22 more are in development. MBDA's missiles will be used on the Eurofighter, Tornado, Rafale, Mirage, Gripen, Tiger and Lynx helicopters and the new European frigates.

Oto Melara became the most important supplier of these weapons due to its reliable small and medium-calibre ship's guns. In addition to these guns, the company also develops solutions for combat vehicle armament. Oto Melara exports to nearly 60 countries around the world.

Italian arms manufacturers have a long tradition in the development and production of handguns, submachine guns and machine guns. The most important company is Beretta, which produces pistols, machine pistols, automatic rifles and machine guns and is also one of the largest exporters of handguns and automatic rifles in the world.

An important supplier of armoured multipurpose vehicles is IVECO, which manufactures special vehicles in Bolzano.

The Italian aviation industry developed large bombers early on

IVECO produces armoured all-purpose vehicles

Aermacchi produces jet trainer (M-346)

The pride of the Italian Navy is the aircraft carrier Cavour

The Dutch shipyards also build warships

The Norwegian company Kongsberg develops simulation solution

4.11. The Netherlands

History

Before the migration of the peoples, the territory of today's Netherlands was populated by the Germanic tribe of the Batavians, Frisians and Saxons. In the 8th century the Franks conquered the area. In the Middle Ages there were several important economic and cultural centres on the territory of the Netherlands. In the 14th and 15th century the area belonged to Burgundy. By marriage the Habsburgs acquired the Netherlands in 1477. The

Netherlands soon rose up against Habsburg rule and a 100-year struggle for freedom began, which ended in 1581 with the separation of the Union states as the Republic of the United Netherlands from Spanish rule. William of Orange was appointed the first governor of the city. The struggle for independence lasted until 1609, when Spain, in an armistice with the Netherlands, also recognised de facto independence. The formal recognition of the independence of the Netherlands took place in the Peace of Westphalia in 1648. In the 17th century, the Netherlands also participated in the colonial policy of Europe. The Netherlands acquired colonies in Southeast Asia, India and America. In the conflicts of the maritime powers, the Netherlands waged several wars against England. In 1689 William III of Orange came to the throne of England. England and the Netherlands were ruled in personal union. After the Napoleonic Wars, Luxembourg, Belgium and the Netherlands joined together to form the united Netherlands. Belgium broke away from this union already in 1830, Luxembourg followed in 1890. During the First World War the Netherlands remained neutral, during the Second World War the Netherlands was occupied by German troops.

Defence Industry

The Netherlands has a large industrial potential with strong innovative capabilities. The relative smallness of the country and the armed forces does not justify an independent larger defence industry, with the exception of naval shipbuilding, so that only in the case of the large civilian companies are parts of the company involved in the development and production of defence equipment.

One of the largest industrial groups in the country, the Stork Fluor Group consists of 69 companies with around 20,000 employees. With the acquisition of Fokker, the company's core competencies also include aviation technology, naval technology, combat vehicles and the construction of building complexes for special tasks. In the field of aviation, technical and logistical support for the maintenance of the F-16 fleet is of importance to the Dutch armed forces. Another aerospace company is Dutch Space, which offers both civil products and intelligent products for military use.

Together with the German companies RheinmetalDec and KrausMaffeiWegmann, the company developed the all-terrain armoured transport vehicle Boxer, a highly mobile 8x8 wheeled vehicle whose modular design allows a wide range of mission-specific vehicle variants.

The main shipbuilding yard in the Netherlands is Damen Schelde Marinebouw in Vlissingen. The naval shipyard has a tradition of more than 125 years and has delivered the most important surface vessels of the Dutch Navy in the last decades, including the frigates of the Tromp, Kortnaer and Karel Doorman class and the landing ship of the Rotterdam class. A new company is the Naval Group Netherlands, which is a subsidiary of the French Naval Group of France.

The Netherlands also has a large industrial potential in the field of communication technology. Globally active companies, such as Thales and Alcatel, also have branches in the Netherlands, which are involved in the development of solutions ranging from simple consumer goods, such as special batteries for portable communication devices, to complex tasks within the framework of the Networked Warfare project. Thales, for example, acquired Hollandse Signaalapparaten, a former foundation of Phillips, in 1990 and produces special batteries, Thales Cryogenics solutions in the field of refrigeration and Thales Communication communication equipment.

4.12. Norway

History

Norway has been Germanically inhabited since prehistoric times. In the 9th century the unification of the country into a unified kingdom under King Harald Schönhaar began. But it disintegrated again very soon. This was used by the Danish kings, who managed to rule the country until 1200. After that Haakon IV. Haakonsson managed to consolidate the royal power from 1217. After the royal house died out, the Danish queen united Norway, Sweden and Denmark in the Union of Kalmar. Norway then remained with Denmark until the Peace of Kiel in 1814. After that Norway had to recognize the union with Sweden. The period of the Norwegian-Swedish union is characterised by constant tensions between the Norwegian Storting (parliament), which was most effective in representing the national and liberal demands for independence and freedom of the country, and the Swedish kings. The tensions eventually led to the declaration of independence in 1905, when the Norwegians decided in a referendum to retain the monarchy and elected the Danish Prince Charles as King Haakon VII. Norway remained neutral during the First World War, but placed its fleet in the service of England. During the Second World War, Norway was occupied and the king and government went into exile in London.

Defence Industry

In the past, the Norwegian defence industry has always been a small part of total industry. This fact has essentially remained until today, so that the Norwegian defence industry develops and produces niche products for the global market.

Norway's flagship defence company is the Kongsberg Group, which generates its sales in the Maritime and Defence & Aerospace divisions. Kongsberg can look back on a long tradition, as the town of Kongsberg was founded in 1624 by the then King Christian IV on the basis of silver findings. After the decline of the silver mine, new business areas were sought and in 1814 an arms factory was founded and the Krag Jorgensen rifle was developed there, which was also exported in large quantities to the USA. Finally in 1849 the shipyard Horten near Oslo was founded. After the Second World War, Kongsberg with its locations in Kongsberg and Horten developed into an economic locomotive, which it has remained despite minor crises in the armaments market.

The Defence & Aerospace division has a long tradition of cooperation with the armed forces and the development of innovative products. Special attention is paid to the development of guided weapons, land and air defence systems, aviation, and the development of communication technologies. Kongsberg products are not only used in the Norwegian Army, but have also been introduced to the Swedish, Romanian, Turkish and Spanish armed forces. Kongsberg's most important products include a weapons control system for combat vehicles, a command and control system for the Norwegian Ula-class submarines, new radios for the army, the NASAMS (Norwegian advanced surface to air missile system) and a new anti-ship missile.

The Norwegian defence industry has the ammunition manufacturer Nammo, which is also very successful in the export business. Nammo Raufoss has its headquarters 120 km north of Oslo and has a long tradition of manufacturing high-tech products. In particular, the company manufactures infantry ammunition from 12.7 mm upwards, anti-tank and artillery ammunition, rocket motors for ammunition and pyrotechnic devices. A special niche product is anti-ship artillery ammunition.

The world's largest supplier of electronic solutions for military purposes, the French company Thales, also has a branch in Norway. With 260 employees, it offers complete solutions for communication and information systems for the Norwegian and Swedish armed forces.

The Norwegian defence industry has a number of companies producing dual-use goods ranging from men's equipment to weatherproof clothing.

4.13. Austria

History

The territory of present-day Austria was initially settled by Celts. Then the Romans conquered the area and established border provinces to Germania. After the collapse of the Western Roman Empire and the beginning of the migration of peoples, Germanic tribes invaded the area. Mainly Bavarian tribes settled there. After a temporary loss to the Magyars and the victory of Otto the Great at the Lechfeld (955) the Mark Bavarian Ostmark was formed, which was called Ostarrichi from 996 on. The Babenbergs were margraves of the Ostmark from 976 to 1246. After the extinction of the Babenbergs, Austria and Styria came temporarily under the rule of the Bohemian King Ottohar II Premysl. When Rudolf of Habsburg defeated Ottokar at Dürnkrut and Jedenspeigen, the area came under the sovereignty of the Habsburgs. This marked the beginning of Austria's rise to become a major European power. The territory of present-day Austria was exposed to constant invasions from the East in the following centuries. Only towards the end of the 15th century were there quieter times until the invasion of the Turks towards the end of the 16th century. The religious battles led to the Thirty Years' War. The power of the empire was reduced as a result. However, the position of Habsburg in Austria was consolidated. After the Thirty Years' War, the balance of power in Europe became increasingly polarised, resulting in almost 200 years of hostility between Austria and France. During this period, France was for the most part an opponent of the Habsburg Empire in all the major wars on the continent (War of Austrian Succession, Wars of the Turks, War of Spanish Succession, Wars of Coalition, Napoleonic Wars, Wars of Italian Liberation). In the middle of the 19th century Prussia won the war against Austria in 1866. This led to the dissolution of the German Confederation. Austria had to resign as hegemonic power in Germany. As a consequence, Austria turned more to the Balkans. With the annexation of Bosnia-Herzegovina in 1908, the tensions between Austria-Hungary and the Slavic peoples of the monarchy and the Balkans, especially Serbia, came to a head. The murder of the heir to the throne Franz Ferdinand on 28 June 1914 led to the outbreak of the First World War. After the defeat in the First World War and the downfall of the Danube Monarchy, a new state emerged in the middle of Europe, which as the Republic of Austria, marked by internal political turmoil, became easy prey for National Socialist Germany. On 12.3.1938 German troops marched into Austria and already on 13.3. the annexation of Austria to the German Reich was completed. Austria, renamed "Ostmark", took part in the Second World War as part of Germany. After the defeat of Germany, Austria was occupied and divided. Due to the Moscow Declaration of 1943, which promised the restoration of an independent state of Austria, the government of Karl Renner was able to proclaim the independence of the Second Republic of Austria on 27.4.1945. However, its powers were severely limited by the four occupying powers until the conclusion of the State Treaty on 15.5.1955. In this treaty the withdrawal of the occupation was agreed between the occupying powers and Austria. Austria was granted sovereignty and economic independence and committed itself to perpetual neutrality. The period after the conclusion of the State Treaty is marked by Austria's accession to the United Nations, the Council of Europe and EFTA. Austria is a member of the EU.

Defence Industry

From the very beginning, the Austrian defence industry had to struggle with the difficulty of a small domestic market and restrictive export regulations, so that this branch of industry always remained a small part of the overall economy in comparison to comparable countries. At its peak in the early 1980s, the Austrian defence industry in the narrower sense employed some 15,000 people. Since the beginning of the 1990s, the Austrian security and defence industry has been in a constant process of shrinking, achieving an annual turnover of € 2.7 billion.

Eight companies produce weapons and ammunition in Austria. The internationally most successful company is Glock GesmbH, which manufactures pistols and personal equipment in Deutsch Wagram. The export share is about 80%. A particularly important market for the company is the United States of America. Another renowned company that produces weapons is the traditional company Steyr Mannlicher AG & CO KG in Steyr. The weapons factory in Steyr is one of the most important and innovative companies in this sector in Europe and produces pistols, sniper rifles, hunting and assault rifles. The products of Steyr&Mannlicher are not only appreciated in Europe, but also in Asia, especially in Malaysia the modern automatic military rifle has been introduced and is also manufactured there under licence. A smaller company that produces weapons for special forces is Voere- Kufsteiner Gerätebau und Handelsgesellschaft mbH, based in Kufstein.

In the field of ammunition production, Austria has the long-established company Hirtenberger Defence Systems GmbH & Co KG, headquartered in Hirtenberg, which manufactures small arms, grenade launchers, tank cannon and artillery ammunition. Another ammunition manufacturer is Armaturen GmbH in Schwanenstadt, which produces hand grenades and especially 40 mm ammunition. Explosives are produced at Austin Powder GmbH in St. Lambrecht in Styria and at Benda-Lutz Werke GmbH in Traismauer.

The Austrian corporate landscape also includes industrial companies that manufacture both military and civilian vehicles and vehicle components. The armaments producer in the narrower sense is Steyr-Daimler-Puch Spezialfahrzeuge AG & Co KG, which belongs to the General Dynamics Group and is based in Vienna-Simmering. The most important products include the "Pandur" wheeled armoured vehicle and the "Ulan" main battle tank. In the past, the company has supplied the German Army with the "Kürassier" fighter tank and the "Greif" armoured recovery vehicle.

The remaining vehicle manufacturers are to be classified in the production group of dual-use goods. The products of KTM Sportmotorcycle AG in Mattinghofen and Rheinmetall MAN Military Vehicles Austria, which manufactures light and medium military trucks in Vienna, are used for military purposes.

The vehicle manufacturing sector also includes companies that produce vehicle components, such as Palfinger Krantechnik GmbH in Bergheim near Salzburg, Pankl, which manufactures chassis and drive systems GmbH in Kapfenberg, Franz Achleitner GmbH & Co KG, which produces low loaders, crane bodies, trailers and special armoured vehicles in Wörgl, and EMPL Fahrzeugwerk GmbH in Kaltenbach, a leading European manufacturer of special vehicle bodies for military use. Another important company producing special vehicles is Rosenbauer International AG, based in Leonding. For decades, Rosenbauer has been regarded as a model Austrian company specializing in the manufacture of fire fighting vehicles, and also supplies many countries with products adapted to military requirements.

The two most important companies producing communication equipment for military use in Austria are Kapsch, the Austrian subsidiary of the international Kapsch Group, and Frequentis. In Vienna, Kapsch produces radios, test equipment, provides comprehensive services and also offers complete solutions for communication, including simulators.

At the beginning of aviation at the turn of the 19th and 20th centuries, Austria - which concerns the development of aircraft - played only a minor role. The development of the "Etrich-Taube" did indeed succeed, but overall the technological yield remained low, so that the Imperial and Royal Family was able to maintain its position in the world of aviation. Monarchy during the First World War was partly dependent on German developments, which were produced under licence in Austrian factories. Total production remained modest, with aircraft produced in total, compared to production in France, Great Britain and the German Reich. In the interwar period and also during the Second World War, the production of aircraft was low. After the Second World War, the resurrected Austria initially had other concerns, so that a modest aviation industry did not emerge until the other more important branches of industry, such as the production of special steels, vehicles, etc., had been established. At the beginning of the 21st century, Austria's aerospace industry therefore presents itself as a small but successful branch of industry that manufactures and distributes high-quality products in close cooperation with foreign partners.

The most important company that manufactures aircraft is Diamond Aircraft Industries. The company is an international, globally operating manufacturer of plastic aircraft, with representatives in Europe, North America, Asia and Australia. At two production sites, one in Wiener Neustadt where the headquarters and the development department are located and one in London, Ontario Canada, innovative aircraft solutions of the highest level and quality are produced for flight schools and private customers. The company employs around 800 people worldwide and has already produced around 3,000 aircraft. The production range includes motor gliders, single and twin-engine piston aircraft and is currently developing single-engine turbine-powered aircraft. The company's reference customers include the French and Indian Air Forces.

Schiebel Elekrtonische Geräte GmbH in Vienna is another important company in aviation in the broader sense of the term, offering helicopter drones in the Unmanned Aerial Vehicle (UAV) segment. The Schiebel company and its product deserve a closer look. The company was founded in 1951 and has been involved in the development of intelligent demining equipment from the very beginning. In the 1990s, the company expanded its fields of business and began work on the development of an Unmanned Aerial Vehicle with vertical landing and take-off characteristics. The result was a mini helicopter Camcopter, weighing around 43 kg with a 25 kg payload capacity. Originally designed for mine detection, this helicopter can be used for many different purposes. The helicopter was able to demonstrate its exceptional capabilities in tests carried out at sea by American government agencies. Wind speeds of up to 21 knots were no problem. A military use seems therefore to be pre-programmed in many ways, for example for radar surveillance, artillery target search, but also for target engagement.

The precision products of Fischer Advanced Composite Components AG are particularly sought after in space technology and also in aircraft construction. Founded in 1989 and based in Ried, Upper Austria, the company produces components for aerospace and aviation with a workforce of around 835. Its customers include all the major aerospace nations and the largest aircraft manufacturers in the world, including Airbus.

The Vienna-based company Photonic Optische Geräte GmbH & Co KG was founded in 1986 as part of the separation of Swarovski Optik into two divisions. The company was acquired by the Wild Group and produces night vision devices, laser rangefinders, aiming devices for artillery guns and grenade launchers and rifle scopes. Another company, Kahles GmbH in Vienna, produces binoculars and rifle scopes. Probably the best known Austrian company in this sector is Swarovski in Absam. Founded in 1905, the company manufactures binoculars, high-precision rifle scopes and laser rangefinders as well as night vision devices.

Compared to other European countries of similar size, population and economic importance, Austria has a modest information technology industry. Nevertheless, there are some companies that develop and produce IT hardware and software for special military applications. For military specific applications Sail Labs Technology AG in Vienna develops special IT software. The company ESL Advanced Information Technology GmbH in Vienna develops solutions for C4ISR (Command, Control, Communication, Intelligence, Surveilance, Reconnaissance) systems and the company AT&S Austria Technologie&Systemtechnik AG. in Vienna is one of the world's largest producers of printed circuit boards. A company that also falls into this section is Geospace Austria, a company that is involved in the production of topographic maps on an electronic basis.

The requirements for field medical equipment for armed forces are based on modern military medical guidelines, which essentially derive their basic principles from disaster medicine. Suppliers of such equipment are the major manufacturers of medical technology, who also maintain an armaments division in order to secure profits in this market segment. The most important companies offering services in this segment are EMPL Fahrzeugwerk GmbH in Kaltenbach, which manufactures medical shelters for motor vehicles,

Goetzloff GmbH in Leonding, which produces equipment for decontamination, Kohlbrat & Bunz GmbH in Radstadt, which manufactures rescue equipment and universal stretchers, Siemens Österreich Medizintechnik in Vienna, which produces medical equipment for a wide range of applications, and the Vamed Group in Vienna, which manufactures medical equipment for field hospitals.

The development of clothing and men's equipment is a part of force planning and includes basic planning and testing through to the preparation of so-called technical specifications as a basis for procurement. The production itself is carried out by companies that usually have the textiles manufactured in low-wage countries in the Far East. In Austria, despite unfavourable market conditions, there are some companies that are involved in the production, but also in the complete in-house development of clothing and men's equipment. Carinthia Gold-Eck GmbH in Seeboden/ Carinthia specialises in the production of high-quality sleeping bags, Gabe Handelsgesellschaft distributes protective vests, combat boots and decontamination equipment, Goetzloff GmbH in Leonding ABC Schutzausrüstungssysteme, The Heat Company in Flachau Schutzsysteme für extreme Temperaturen, Sattler AG in Graz Zeltsysteme und persönliche Schutzausrüstung and the company Glock mentioned in the first part of the attachment series, which also sells men's equipment in addition to weapons.

The picture shows products from the Austrian defence industry: KTM wheels, Pandur wheeled armoured vehicles, lifting cranes, assault rifles, pistols and special ammunition

The picture shows products of the Austrian armaments industry: Diamond aircraft, Schiebel drone, mine detectors, fighter command vehicles, sniper rifles and battle tanks

4.13. Poland

History

Poland has a long history, which began in the 6th and 7th centuries with the immigration of Slavic tribes into the territory of today's Poland. Already around the middle of the 9th century, the unification of the tribes and the establishment of a central Polish territory under the House of Piasts succeeded. Very soon, however, the empire disintegrated into a number of independent duchies. As a result, the individual parts of the former Central Polish territory became easy prey for the German settlers pushing from west to east. It was only under the Jagiellons that reunification was achieved around 1386. After the extinction of the Jagiellonians, an elective kingship was established, with foreign princes occupying the Polish throne. Different interests of the high nobility and the influence of European powers led to the decline of Poland around the middle of the 18th century and to the first partition of Poland around 1772. A second partition followed in 1793 and finally Poland was divided for the third time in 1795. Although Napoleon I created the Grand Duchy of Warsaw and the Congress of Vienna created the Kingdom of Poland, which was personally united with Russia, Poland remained dependent on the European powers. Only after the First World War was an independent Republic of Poland created, which received its present form after a territorial correction after the Second World War. After a period of communist rule, a change occurred and Poland today has a democratic form of government. Poland is a member of the EU and NATO.

Defence Industry

At the end of the 1980s, the Polish defence industry employed some 215,000 people in 150 companies; by 2003, the figure had fallen to 50,000 in some 58 companies. The Polish defence industry has always strived to cover a wide range of products. These range from battle tanks and armoured vehicles, artillery and anti-aircraft weapon systems, aircraft and helicopters, naval vessels and ordnance, certain electronic equipment and communications equipment to light weapons and ammunition.

Polska Grupa Zbrojeniowa (Polish Armaments Group) is the market leader in the production of main battle tanks, armoured vehicles, artillery pieces and small arms and machine guns. The holding company consists of about 15 companies that produce all kinds of armoured vehicles, including licensed products of the Finnish Patria Group.

The most important company in the aviation sector is PZL in Swidnik, which belongs to the Italian Leonardo Group and produces helicopters, and Polskie Zaklady Lotnicze - PZL Mielec, which belongs to Sikorky and produces aircraft and aircraft parts.

The electronics sector is dominated by the companies Radmor, WB Electronics and the Wojskowy Instytut Lacznosci (Military Communications Institute). Radmor is a specialist company for the development of telecommunication equipment, night vision technology and fire control systems. Radmor also cooperates successfully with US defence company Lockheed Martin.

Naval shipbuilding is mainly carried out at Damen shipyard in Gdynia. This shipyard produces landing ships, guard boats and meko frigates.

4.15. Portugal

History

Before our era, the land between the Douro and the Algarve coast was inhabited by a Celtic-Iberian mixed population. Only in the south there were some Carthaginian colonies. Despite fierce resistance the Romans established a province of Lusitania. During the migration of the peoples,

Visigoths and other Germanic tribes settled there. The Visigoth empire was destroyed by the Arabs in 71 AD and the land was incorporated into the Saracen dominion. The reconquest began in the 9th century. The reconquest was completed around the middle of the 13th century with the conquest of the Algarve and the border with Spain, which is still valid today, was established. In the following years, the Portuguese took advantage of the flourishing European trade and Lisbon became one of the most important sea and trading cities in Europe. At the beginning of the 15th century, Portugal laid the foundations for the Portuguese colonial empire by conquering the trading and pirate centre of Ceuta in Morocco. In the following centuries, the Portuguese empire was gradually expanded. It was not until the first half of the 19th century that an important cornerstone of the empire became independent. In 1822 Brazil became independent. After that, Portugal did not come to rest. This was followed by internal unrest, which led to the abolition of the monarchy and, over a short phase of a republic, to a dictatorship. It was not until the mid-1970s that the dictatorship was abolished and democracy returned.

Defence Industry

Due to its relative smallness, the Portuguese armed forces require only a small amount of defence equipment, which has not yet justified the establishment of an efficient defence industry on a national level. The Portuguese defence industry is accordingly kept small and is grouped together in the holding company Empresa Portugesa de Defesa EMPORDEF), which is state-owned. The most important parts of this holding company are two shipyards, Lisnave Estaleiros Navais and Arsenal do Alfeite, two ammunition factories, an aeronautical engineering company, Ogma-Industria Aeronautica de Portugal S.A. and two companies in the electronics sector, EdiSoft-Empresa de Sevicos e Desenvolvimento de

Software S.A. and EID-Empresa de Investigacao e Desentvolvimento
Electronico S.A.

The Lisnave Estaleiros Navais shipyard was founded in 1937 and built some
frigates for the Portuguese navy in the early 1960s, but since then it has been
doing the main business of maintenance and modernization for the navy and
also carries out repairs and modernization. The second shipyard, the Arsenal
do Alfeite, has so far built watchboats and landing craft for the Portuguese
Navy.

In the aviation industry, OGMA is involved in the maintenance and
modernization of military aircraft and ground facilities as part of its business
activities. It also operates a development office for new products and system
solutions for the military aviation sector.

In addition, Portugal has a number of electronics companies that provide
special solutions to tasks from the Ministry of Defence.

4.16. Romania

History

The territory of present-day Romania came under Roman rule in 107 AD and
was rapidly Romanised. After the invasion of Germanic tribes during the
migration of peoples, Slavs began to invade the area from the 6th century
onwards. The principalities of Moldavia and Wallachia were formed, which
came under Ottoman rule in the 15th century. The independent kingdom of
Romania is a birth of the Berlin Congress of 1878. Romania experienced its
greatest territorial expansion after the First World War when the Trianon
Peace Treaty granted Bessarabia, the Eastern Banat, Bucovina and
Transylvania. After the Second World War, Romania came under Soviet
influence and, moreover, the Peace of Paris finally forced it to give up some

territories it had de facto already ceded to the Soviet Union and Bulgaria in 1940. These were North Bucovina and Bessarabia to the USSR and South Dobruja to Bulgaria. Despite its membership in the Warsaw Pact, Romania pursued a relatively independent foreign policy. In Romania, the last domino in the collapse of the external empire of the Soviet Union, a violent, bloody overthrow of the old regime took place. The revolutionary uprising began in 1989 and since then the country has been undergoing a process of democratisation. Immediately after coming to power, the new Romanian leadership sought to restore the consensus with the other Warsaw Pact states, and especially with the Soviet Union, which had been disrupted since the 1960s. After the fall of the Iron Curtain, the government in Bucharest had also attempted to revive the preferential relations with the European Community that had existed since the 1970s. The initial ambivalence of Romanian foreign policy between 1989 and 1991 put a strain on the country's relations with the states of Western Europe.

Defence Industry

Before 1990, some 80,000 employees in some 30 companies in the defence industry produced military equipment, which led to sales of around 1 billion US dollars. From 2001 onwards, a massive downsizing process set in, which led to a sharp reduction in jobs, with the result that the Romanian defence industry currently employs around 18,000 people and generates sales of only around 30 million US dollars.

In the aviation industry, there are two state-owned companies, Intreprinderes Aeoronautica Romana SA (IAR) and Avione SA, and two partly privatised companies, Aerostar SA and Romaero SA. IAR, for example, produces under license the Eurocopter IAR-316B, the Alouette III, IAR-330L and the Kamov Ka-126. The Romanian company Aerostar, which specializes in the modernization of fighter aircraft, also has a joint venture with the Israeli defense company Elbit.

Land systems are manufactured by ROMARM SA. This holding, consisting of 15 companies, currently develops or produces weapons and weapon systems.

Marine systems are repaired in the shipyards of Constanta and Galati.

4.17. Russian Federation

History

The multi-ethnic state, which extends over two continents - Europe and Asia - was founded by the Kiev Grand Duke Vladimir I, the Saint (978-1015). Under the Mongol storms of 1223, 1237/38 and 1239/40 the Grand Duchy broke up and the partial principalities came under the sovereignty of the Western Mongolenkhanate, the "Golden Horde" with its seat in Sarai on the lower Volga for a century and a half.

Only under Ivan III. (1462-1505) there was a rise again and one generation later Ivan IV. (the Terrible, 1547-1584) was crowned "Tsar and autocrat of the whole great Russia" - a title that included the claim to the succession of the Byzantine imperial dignity. Under his reign, the empire was expanded to the east and south in the context of imperial power politics. Internally, a phase of reforms was followed by the Regiment of Terror, the Opritschnina. With his son Fjodor, who died in 1598 without a successor, the dynasty of the Rjurikids came to an end. After a phase of "smuta", the turmoil, a new beginning was made in 1613 with the election of the Boyar Mikhail Romanov (1613-1645).

The boyars were members of the high nobility in ancient Russian history. It was then his descendant Peter I (1682/1689-1725) who modernised Russia with a firm hand and opened the "Gates to Europe". Tsar Peter built up a seaworthy fleet, enlarged the army and trained it according to Western models. He reorganized the central administration by creating colleges (ministries) structured according to specialist tasks following the Swedish model, and he pursued a mercantilist economic policy by promoting trade and domestic production through the establishment of manufactories. Foreign specialists were brought into the country, Russians were sent abroad to study. On January 1, 1700, the Tsar also had the old Byzantine calendar abolished, which was replaced by the Julian calendar following the example of the Protestant countries. (The more accurate Gregorian calendar, which was valid in the Catholic regions and to which we still adhere today, was not introduced in Russia until 14 February 1918).

But large sections of society, especially the peasantry, remained unaffected by the Petrine attempt at modernization. The contradiction between the tradition-oriented majority of the population and the modern absolutist regime determined Russia's social and intellectual life well into the 20th century. But Peter's reforms created the conditions for the country's rise to become a major European power. After the victory over the Swedes in 1709 at the Battle of Poltava (now in Ukraine), the Tsar accepted the title of Emperor, which was immediately recognised by Prussia and the general states of the Netherlands, and by most other European powers over the next 50 years.

In the 18th and 19th centuries, Russia developed into an important player in European politics. Under Peter's successors, the empire expanded to the south in wars with the Ottoman Empire and to the west through the Polish divisions. In the Napoleonic Wars at the beginning of the 19th century, the Tsarist Empire played a decisive role as an opponent of Napoleonic France, then as its ally and finally again as a member of the anti-Napoleonic alliance, which also played a decisive role in the reorganization of Europe at the Congress of Vienna in 1815. In the years that followed, Russia joined forces closely with the conservative powers Habsburg and Prussia within the framework of the "Holy Alliance" and, as the "gendarme of Europe", attained a decisive role on the continent. The tsarist government pursued a restorative-conservative policy and turned against all freedom movements of the time. In 1861 serfdom was abolished under Tsar Alexander II. Russia took part in the First World War on the side of the Entente and in 1917 there was a democratic revolution in Russia. Tsar Nicholas II had to abdicate. The Bolsheviks, led by Lenin, took power. The tsar's family was imprisoned and murdered by the Bolsheviks in 1918. Josef Vissarionovich Dschugashvili, called Stalin, came to power in 1924. During the "Great Purge", Russian "Chistka", 1936-38 he liquidates almost all his rivals. In 1941 German troops attack the USSR without a declaration of war and the "Great Patriotic War" breaks out. On 9 May 1945 the USSR triumphs over Hitler's fascism. Stalin dies in 1953. After his death democratization, turning away from the personality cult and a liberalization of the judiciary set in. With the era of Mikhail Gorbachev. The socialist system is restructured and reformed. Despite controversial economic policy, his merit is immense: he opens the country to the outside world and introduces democratic freedoms. The terms "glasnost" and "perestroika", which he coined, are understood throughout the world. Boris Yeltsin, Russia's first democratically elected president, introduces market economy and "presidential democracy" in 1991. 2000 Vladimir Putin becomes the strong man in Russia.

Brief history of the Russian armaments industry from the 18th century
Tsarist Russia had a number of armament factories, which were founded towards the end of the 18th century. These included the Putilov Works, the Obuchov Works and the Tula, Izhevsk and Sestroretsk arsenals. The Putilov Works were founded in 1789 as a state ironworks near St. Petersburg. The

plant was named after Nikolai Ivanovich Putilov, an employee of the Navy Ministry. The plant experienced its greatest expansion between 1885 and 1900, when it started producing guns and armoured road vehicles. The arsenals in Tula, Ishevsk and Sestroretsk were designed as cities. They contained schools, military hospitals, department stores and also training facilities. The first arms factory in Tula was built as early as 1632 by the Dutchman Franz Marcellus. In 1712, the imperial arsenal was established by a decree of Tsar Peter the Great. At the end of the 19th century, around 1882, production was greatly increased, around 5,000 additional workers were hired and almost 160,000 Berdan rifles were produced. In the last years of the 19th century the machinery was renewed and various military rifles were produced. Ishevsk, which lies 77 km northwest of Sarasul on the river Ischa at the foot of the Ural Mountains, was founded in 1760 as an ironworks and has been producing weapons since then. The arsenal itself was only founded in 1807. Between 1807 and 1907 the Arsenal produced about 4 million rifles. The factory exists today as Izhmash, a weapons factory mainly for sporting weapons. Weapons production in Sestroretsk began around 1724. The factory was built about 27 km northwest of St. Petersburg on the Kronstadt hill near the Gulf of Finland and, under its managing director Colonel Mosin, produced the "M 1891 three-line rifle" from 1894 onwards. Dreiliniengewehr was a designation of the caliber specification in the old unit Linie, where 3 lines correspond exactly to 7.62 mm. The workers of the Sestroretsk Arsenal played during the October Revolution, as the Arsenal was located near the Lenin headquarters. During the Civil War the Arsenal was evacuated because the Leninist troops were afraid that the White Army could take the factory. The factory no longer exists today.

The Russian Federation was the main heir to the arms industry of the former Soviet Union. The territory of today's Russian Federation was then home to the most important and largest arms companies (around 75 percent), which produced for the then great empire and also for the high level of exports to friendly states with unrestricted access to all state resources.

Naval shipbuilding industry

In the fleet lists of the Russian fleet, the first warship to appear is the sailing ship Oryol. The ship was laid down on 14 November 1667 in Dedinovo on the left bank of the Oka near the confluence of the Moskva and was launched on 25 May 1668. With the Oryol's entry into service the building of the war fleet in Russia began. But it was the young Tsar Peter I who in 1696 urged the Duma to decide that Russia should build sea-going ships. Thus, by the end of 1698 it was planned to complete 77 warships, 609 river ships, 509 boats and 400 rafts. However, these grandiose plans met with resistance from the church, the merchants and some of the boyars. The general conditions for warship building in Russia at that time were very bad. On an inspection of the ships already built, Peter I had to admit that of fifteen ships only nine were actually ready for action, but they too still needed major changes. After this failure, Peter I himself learnt the shipbuilding trade abroad and took well-meaning comrades-in-arms into Russian service. Peter I. travelled to Western Europe, especially to Holland and England, and acquired practical knowledge of shipbuilding. He also acquired 22 guns, 5000 rifles, 3200 bayonets and other instruments. He also hired about 1000 shipbuilding specialists. After his return, Peter I resolutely set about implementing his plan to create a Russian navy. In his presence a warship armed with 60 guns - the Goto Predestinazija ("God's Foresight") with a length of 36 m was stacked in Voronezh. At the same time as the decision was made to build a war fleet, it was also decided to build a merchant fleet in Arkhangelsk. This caused unrest in Europe. Especially the Dutch were confronted with losses for their own merchant fleet. The fears were unfounded at first, however, as Russia was prevented from building a merchant fleet by the ongoing war against Sweden. However, the shipyard of Arkhangelsk became bigger and more efficient. This circumstance also allowed the construction of liner ships in Arkhangelsk. In 1702 another shipyard was built in Novgorod, it had to stack 6 frigates and 10 smaller ships according to the decree of the tsar. One year later, in 1703, another shipyard was opened in Lodeinoje Pole on the bank of the Swir. This shipyard was later given the name Olonezk shipyard. Until the middle of 1703, 7 frigates, 5 pinass ships, 7 galleys, 13 semi-galleys, 1 galiot and 13 brigantines were stacked at this shipyard. However, these shipyards had the disadvantage that the ships built were difficult to bring across Lake Ladoga to the Neva. Peter I therefore decided that a new large arsenal (admiralty) should be built on the banks of the Neva. The construction of the shipyard began in 1705 and in 1706 the first Pinass ship, the Nadeshda, left the shipyard. Peter I himself tested the ship by taking the Nadeshda to Kotlin Island. The construction of large warships at the shipyard did not begin until the end of 1709, however, when Peter I himself constructed the Poltava, a warship for the open sea armed with 54 guns. The ship was launched in 1712. The ship took part in all important campaigns of the Nordic War. At the shipyards of the Admiralty the speed of shipbuilding was very high. In the course of 1712 alone, 6 liner ships and 50 smaller ships were delivered. In total, 20 liner ships and frigates, 15 pinasses, 2 bombarde ships, 4 prams and 170 smaller ships were built at Russian shipyards between 1703 and 1711. However, lack of experience meant that only some of the ships built were actually able to take part in the battle. At the end of Peter I's reign the Russian fleet consisted of 48 liner ships and frigates, 787 galleys and other ships and 28,000 men. After the death of Peter I the fleet decayed. Only with the Tsarina Catherine II. the navy moved again into the interest of the planners. A new naval shipyard was built in Kherson by tsarist decree. In 1788 a new shipyard was built at the confluence of the Ingul and the southern bow on the order of the Russian General Potjomkin. The shipyard was named Nikolayev in honour of Saint Nikolay. It became one of the most important shipyards in Russia. In the middle of the 19th century the revolution of the steamships began. The Russian shipyards Arkhangelsk, Astrakhan, St. Petersburg, Nikolayev, Alexandrovski were to implement the shipbuilding programme for iron ships decided by the Russian government from 1857. All shipyards had to be modernised for this purpose. A two-stage plan was envisaged for the construction of a fleet of ironclad ships. First a fleet for coastal protection was planned, and then, at the end of the sixties, a deep-sea fleet. In the course of the years 1870 to 1885 Russia completed its deep-sea fleet. At the beginning of the 1880s, a new shipbuilding programme was presented, which covered a period of 20 years. As relations with Japan became increasingly strained at the beginning of the 20th century, a shipbuilding programme called "Necessities for the Far East" was launched. Russia had to build its fleet in the Far East without shipyards and docks, undoubtedly a huge disadvantage compared to Japan, which was able to build its fleet under the protection of the alliance with Great Britain.

During the period of the Cold War, the Soviet Union managed to build up an efficient navy. After the collapse of the Soviet Union, the Navy also collapsed.

Only from 2005 onwards, a new beginning is discernible and the yards start producing new battleships again. The United Shipbuilding Corporation (USC) is an association of 40 shipyards on the Baltic Sea, the North Sea, the Black Sea and the Caspian Sea. The company employs around 95,000 people and produces all types of combat ships. In the east of the country there is an independent shipyard in the Vladivostok rope.

Military Aviation Industry

The beginnings of Russian military aviation date back to the 1880s, when Russian scientists Nikolai Kibalchich and Alexander Mozhaisky were researching theoretical projects related to aviation. During the 1890s

Konstantin Tsiolkovsky further developed the theories. Based on these theories, Nikolai Zhukovsky founded the first aerodynamics institute in Kachino near Moscow in 1904. In 1910 the Russian armed forces bought aircraft from France and began training their military pilots. At the beginning of the war, the tsarist air force had 360 planes, and another 353 were ordered in France. The history of the Russian aviation industry is closely connected with the name Igor Sikorsky. In 1913 Sikorsky built his first aircraft, the four-engined bomber "Russkij Witiaz". It was the first four-engined bomber in the world. His masterpiece became the bomber "Ilya Muromets", a four-engined bomber. The bomber was produced in series by the Russian-Baltic Wagon Factory, so that a total of about 80 pieces could be built. The bomber was successfully used against German and Austro-Hungarian war important facilities. The first mission was against railway installations and sea-flying stations in East Prussia. During an air raid, the Eskadra Wozduschnich KJorablei (EWK) - the squadron of flying ships - destroyed the headquarters of the German commander-in-chief in Lithuania. The planes were extremely robust. The German air defence succeeded in shooting down only one Ilya Muromets. The success of the giant aircraft aroused the interest of the French and British. Licensed production had already been approved by Tsar Nicholas II, but production in England and France was no longer possible, as England had also produced a bomber in the meantime. Between 1914 and 1917 Russia built about 5,000 aircraft, comparatively little compared to the German Reich, which built about 45,000 aircraft during the First World War. After the Soviets came to power, Sikorsky left the country in 1919 and made his brilliant ideas available to the United States of America.

After the consolidation of power, the conditions for the Luftwaffe also began to consolidate in the mid-1920s. From that time on, the Red Air Force started to equip its air force with newer foreign and newer Russian equipment. In the five-year plans that followed from the mid-1920s onwards, the Russian aircraft industry expanded. The design offices and manufacturing plants of Antonov, Beriev, Ilyushin, Yakovlev, Lavotschkin, Lissunov, Mikoyan/Gurevich, Petlyakov, Polikarpov, Sukhoi, Cheverikov, Tupolev were established. The Russian aircraft industry produced around 40,000 aircraft during the Second World War. The best known aircraft of the Red Air Force at the beginning of the war were the fighters and fighter-bombers Lawotschkin LaGG-3 (6,258 pieces), Mikoyan/Gurevich MiG-3 (3,120 pieces), Polikarpov I-16 "Rata" (9,000 pieces), Ilyushin Il-2 "Sturmovik" (Different sources give production figures from 29,937 pieces to 36.163 pieces) as well as Yakovlev Yak-1 and the bombers Petlyakov Pe-2 (11,427) and Pe-8. At the end of the war the fighters and fighter-bombers Mikoyan/Gurevich MiG-5, Lawotschkin La-5 (9.920 pieces), Yakovlev Yak-9 (16,769 pieces) as well as Lavotschkin La-7 and the medium-weight bombers Ilyushin Il-4 (5,256 pieces) and Tupolev SB-2 "Katyushka" (6,656 pieces).

Even before the end of the Second World War the Soviet Union tried to build a jet fighter. The design offices of Mikojan/Gurewitsch, Jakowlew, Lawotschkin and Suchoj were commissioned to design and build a jet fighter. However, the first prototype of a Russian jet fighter did not fly until 1946, and almost at the same time the design offices of Mikojan/Gurewitsch and Jakowlew succeeded in flying a jet fighter. The MiG-9 as the prototype was later called had BMW engines installed. About 1,000 copies of the aircraft were produced. For the Yak-15 the designers used a Junkers Jumo 004 jet engine which was also used in the Messerschmitt Me-262 and in the Arado Ar 234. About 280 of the Jak-15 were produced. The aircraft was mainly used for training purposes. From 1948 on the office Lawatschkin succeeded with the production of the La-15 in producing a jet-powered aircraft. It was very popular with pilots but production was stopped after 500 copies to free all capacities for building the MiG-15.

The MiG-15 was an extremely successful fighter aircraft which was flown by the USSR and all brother air forces from 1949 on. During the Korean War the aircraft proved to be superior. Only the North American F-86 proved to be equal. About 8,000 of the MiG-15 were produced in the USSR, CSSR, Poland and the People's Republic of China. With the MiG-15 the success story of the Mikoyan/Gurevich design office of the post-war period began. Until today the office produced more than 50,000 fighter aircraft of all types, among them such successful models as the MiG-19, the first Soviet supersonic combat aircraft, the MiG-21, probably the most built combat jet of the post-war period, the MiG-25, a supersonic interceptor and the MiG-29.

After the October Revolution and the takeover by the Bolsheviks, a number of aircraft manufacturers were established, which will be briefly described below.

Sukhoi was founded in 1939 by Pavel Sukhoi. The production facilities were located in Siberia from the beginning. During the Second World War Sukhoi built a light bomber. After the war, the company turned to the construction of jet-powered fighter planes. Remarkable was the construction of the Su-9, a jet that looked similar to the Me-262 and had its maiden flight in 1946. It served as a base for the Su-11.

Tupolev was founded in 1922 by Andrei Nikolaiyevich Tupolev. The company specialized in the construction of bombers. The most important types used during the Second World War and in the post-war period were Tu-2, Tu-4 and Tu-16. The design of the Tu-4 was remarkable. For the design, Boeing B-29 Superfortress were evaluated, which had landed in Russia in 1945 after a mission in Japan.

Yakovlev was founded in 1932 by Alexander Sergeyevich Yakovlev in Moscow. Yakovlev built a number of successful fighter planes during the Second World War. Yakovlev also tried his hand at designing helicopters and vertical take-off aircraft, including the Yak-36 and Yak-38 models, which were used on the Kiev class aircraft carriers.

Antonov was founded as aircraft factory No. 153 in Novosibirsk in 1946 and started to build aircraft for the agricultural industry. The first aircraft was the popular AN-2, after which the factory moved to Kiev in 1952 and built mainly transport aircraft for civil and military use. After the death of the chief designer O.K.Antonov in 1984, it was renamed Antonov Aircraft Works in his honour.

Beriev was built in 1934 by Georgy Mikhailovich Beriev in Taganroy and specialized in the construction of amphibious aircraft. Until today, they built about 20 different models for civil and military use.

Myasischev was founded in 1951 by Vladimir Myasishev and specialized in the production of long-range bombers and high-flying reconnaissance aircraft.

Ilyushin was founded in 1933 by Sergei Vladimirovich Ilyushin. The company developed and produced highly successful fighter aircraft, including the Il-2, and also transport aircraft, including the Il-76.

The development and construction of helicopters began in Tsarist Russia with the attempts of Igor Sikorsky around 1909/10, but Sikorsky did not succeed in constructing a helicopter that could take off with a pilot. Discouraged by this, Sikorsky turned to the construction of fixed-wing aircraft and designed one of the best bombers of the First World War, the

"Ilya Muromets". However, Sikorsky was not the only Russian who experimented with helicopters. K.A. Antonov built a helicopter plan in 1910 and in 1911 Boris N. Yuryev built an extremely remarkable construction of which we know too little to this day. We only know that the construction had a vertical rotor and a tail rotor. However, we do not know whether the machine ever flew. After the October Revolution and the consolidation of power, the new government decided to build its own national aviation industry. In 1925 a department of rotor aircraft was formed in the already established Central Air and Hydrodynamic Institute (TsAGI) in Moscow. One of its first members was Jurijew. The first successful aircraft was not a helicopter, but a rotorcraft similar to the Cierva types. It was designed by Nikolai Kamov and N.K.Skrschinski and was named KaSkr-1. After the war it was mainly Michael Mil who made the first successful helicopter designs. The first successful design was the type GM-1, which flew as a prototype in 1947 and finally went into series production in 1951. The model was exported to 17 countries. In 1955 the program was transferred to Poland and continued to be built there until 1964. In 1955 Mil started the development of the Mi-4, a larger helicopter whose design was obviously inspired by the Sikorsky Type 55. Besides Mil, Yakovlev also tried his hand at helicopter design. The Yak-24 flew for the first time in 1953 and was put into service with the Soviet Army in 1958. It was the first large transport helicopter of the USSR, it could transport up to 30 people. But Kamov also took part in the race for the largest helicopter. His Ka-15 flew in 1952 and was used in large numbers, especially in the Navy. Without going into further detail, Mil and Kamow developed a series of powerful helicopters.

After the end of the Cold War, the fragmented structure of the production of fighter planes and helicopters was reorganized. Today there are mergers of Russian Helicopters, United Aircraft Corporation and United Engine Building Corporation.

The large enterprise Russian Helicopters was formed by the helicopter manufacturers Mil and Kamov. It employs about 46,000 workers.

The major company United Aircraft Corporation combines the aircraft factories of Ilyushin, Irkut, Mikoyan, Sukhoi, Yakovlev and Tupolev. It employs around 102,000 people.

The United Engine Building Corporation is a company that includes the major engine manufacturers.

Military vehicle industry

Tsarist Russia had no significant automobile production. The most important manufacturers of cars and trucks were the Russian-Baltic Wagon Factory in Riga, founded in 1874, and the Avtomobilnoe Moskovskoe Obshchestvo (AMO), founded in 1916. From 1909 to 1915, the Russian-Baltic Wagon Factory built automobiles and a truck with a payload of 5 tons for the armed forces. AMO acquired the license rights to build a replica of the FIAT F-15 1.5 tonne lorry. The production of the AMO-15 started only in 1924 due to the turmoil of the October Revolution and the following civil war. In 1931 the AMOS-3 was built, a more powerful truck based on an American type. In 1932, the name of the company was changed to Zavod Imieni Stalina (ZIS) and later in the Khrustchov era to Zavod Imieni Likhacheva, a director of the plants (ZIL). From 1934, the ZIS company built the ZIS-5, which was used in large numbers by the Republican troops during the Spanish Civil War. The successor model, the ZIS-6, a 6x4 truck also served as a chassis for the notorious "Stalin organ", a Soviet multiple rocket launcher for unguided missiles. Besides trucks, ZIS also built two types of half-track vehicles, the ZIS-33 and the ZIS-42. Heavy trucks were manufactured by Yaroslavl Automobilni Zavod). In the factory in Yaroslavl on the Volga, trucks of the types YA 3 and YA 4 were produced from 1924, and in the 1930's the types YA 10 and YA 12. The YA 12 was an 8x8 truck with a payload of 12 tons. Another important automobile plant is Gorkovsky Avtomobilny Zavog (GAZ), which was founded in 1929. The GAZ company was built on an area of 256 hectares and employed up to 12,000 workers. GAZ produced about 42,000 GAZ-A type staff vehicles. Other light staff vehicles were types GAZ 11-40, GAZ-61 and GAZ-67. GAZ was also an important supplier of trucks. The most important models were the GAZ-AA and GAZ-MM and GAZ-AAA. Between 1931 and 1940, the total production of buses and trucks in the Soviet Union was about 1,020,900 units, of which GAZ is credited with about 70%. The Soviet Union also produced artillery tractors, such as the Stalinets-65 and the somewhat smaller STZ5-2B.

Today, military vehicle production is dominated by the Kamaz company, which manufactures civilian and military vehicles with around 36,000 employees.

Production of heavy weapons and ammunition
From 1991 to about 1997 the production volume of the Russian arms industry fell by almost 90 percent. The sharp drop in production led to massive job cuts. The number of employees fell from 10 million during the Soviet era to less than 2 million in 2000. In 1998, the trend was reversed and growth in the production of military equipment was noted again.

When Vladimir Putin took over the office of President of the Republic, the restructuring of the ailing arms industry was once again raised as an issue. In 2002, the Putin administration enacted the "Armament Program 2002-2010", which sets priorities for research and development in the period 2002 to 2005 and for the production of the developed weapon systems until the end of the decade. An important goal in this context is also the restructuring of the former military-industrial complex. To this end, the Programme on Weapons Production in State Ownership was adopted for the years 2001 to 2010. One of the most important objectives of this measure is the better exploitation of existing resources and also the use of research results in the military goods sector also for civilian production. First progress of this bundle of measures is already visible; the defence industry of the Russian Federation is again represented on the global defence market as an exporter and can already make profits again. The export of military equipment has become a success again, not least because a new export agency, Rosoboroneksport, has succeeded in making export initiation and processing more efficient. In the following, the most important companies of the new armaments industry will be presented.

The largest company producing land systems is Almaz-Antej, which became the largest producer of weapons in the Russian Federation through the merger of about 45 companies. The most important product range includes air defence systems, for example complete solutions in the SAM range.

The Military Industrial Company with almost 6,000 employees produces BTR-82A armored personnel carrier, VPK-233136, VPK-3924, VPK-3927 Volk and the VPK-7829 boomerang.

Uralvagonzavod in Niznij Tagil is important for the production of heavy weapons and employs about 30,000 people. The traditional company, which can look back on a history dating back to 1931, produces military equipment with about 30 per cent of its capacity, the remaining 70 per cent being available for the production of civil goods. The plant produces battle tanks, armored vehicles.

The Kalashnikov company, which employs 6,000 people and produces small arms and machine guns, is a major player in the production of small arms.

Important in the field of electronics manufacturing are the companies Ruselectronics with 80,000 employees, Shvabe with 18,000 employees and Radio Electronic Technologies Concern (KRET) with 50,000 employees.

There are also a number of companies in the Russian Federation that manufacture men's equipment and shrapnel protection systems.

The Russian industry is capable of producing high-quality combat aircraft.

The navy has high-quality combat ships from domestic production.

**The Romanian defence industry produces weapons for all branches of
the armed forces**

The Portuguese defence industry produced wheeled tanks

The pride of the Russian submarine fleet are the strategic submarines of the Typhoon class

The Polish defence industry produces helicopters of various load classes

4.18. Sweden

History

In prehistoric times, southern and central Sweden was settled by Germanic tribes. In the 9th and 10th centuries a number of small kingdoms were established on the territory of present-day Sweden. Only for a short time in the 12th century, the Yngling family was able to establish a unified dominion in Sweden. After that the ruling house came under increasing pressure from the powerful nobility. Towards the end of the 14th century, around 1384, it came to the Union of Kalmar, in which the Danish Queen Margaret I. also became Queen of Sweden. Strong nationalist currents caused the Union of Kalmar to break up. In 1527 the rise of the royal house Wasa followed. With the establishment of the hereditary monarchy and the Reformation in Sweden, a phase of expansion began, which initially brought Sweden more territory on the Baltic coast. Only the defeats of Charles XII in the Nordic War of 1700-1721 put an end to the Swedish advance and at the same time meant the return to the original territory. Through the defeat, Sweden lost a large part of its possessions outside the motherland. Another defeat during the wars against Napoleon I led to the loss of all areas outside the motherland. The king was then overthrown and the Bernadotte dynasty was established, which still rules today. In the peace of Kiel in 1814 Sweden won Norway, which administered it in personal union until 1905. Sweden remained neutral during the two world wars.

Defence Industry

Sweden is one of those European countries with a very eventful war history. For years they waged war against Denmark and also participated successfully in the Thirty Years' War. Only the defeat in the Nordic war against Russia led to a rethinking in Sweden. Finally, it was decided to pursue a policy of neutrality in the future, which was also adopted during the First and Second World Wars. During the wars as well as during the period of neutrality, Sweden tried to achieve independence from foreign defence industries by producing weapons in its own factories. Thus, already in the 16th century, warships were launched in own shipyards; the warship VASA, even if the sinking became a mockery of all people immediately after the launch, testifies to the great shipbuilding art of Swedish craftsmen. At the latest since the time of the Thirty Years' War there was also an own production of powder, small arms, rifles and artillery guns. Some of the companies founded at that time are still active in the armaments business today.

The Swedish defence industry, like the defence industries of other countries, suffered from the fall of the Iron Curtain and the resulting reduction of the armed forces. The change in defence requirements and the associated reduction in production led to a reduction in the number of people employed in the defence industry in Sweden from 22,700 in 1990 to 14,400 in 2001.

Sweden has some 25 companies with a strong defence industry. A large proportion of these enterprises are already foreign-owned or have a high foreign ownership share. The top 8 companies, Alvis Hägglund AB, Bofors Defence AB, Ericsson Microwave Systems AB, Eurenco Bofors AB, Kockums AB, Nammo Sweden AB, SAAB AB and Volvo Aero Corporation will be briefly described in this paper; the others will only be mentioned by name with their product range.

Sweden has a great tradition in the production of combat vehicles. Innovative solutions have always been on the agenda, such as the Stridsvagn. The largest supplier of land systems is Alvis Hägglunds, a company founded in 1899 and owned by the British defence contractor BAE Systems. At the Örnsköldsvik production site, the CV90 main battle tank is currently being manufactured.

Ammunition production has been a tradition in Sweden at least since the time of King Gustav II Adolf, when a powder factory was established in Sweden to supply Swedish troops during the Thirty Years' War. Nammo Sweden is part of the Northern European ammunition manufacturers' network and produces ammunition for small arms, battle tanks, artillery and pyrotechnic explosives at its production facilities in Karlsborg, Gothenburg and Karlskoga, Lindesberg and Vingoek. Another supplier of ammunition is Eurenco Bofors, which produces explosives and pyrotechnic agents.

In the field of electronics, Sweden has Saab/Ericsson, which develops and produces radar systems for land and air forces. These include Erieye, an airborne early warning system; Arthur, an artillery hunting radar; Giraffe, a 3D radar for the land forces. The company also supplies the electronics for the Swedish Gripen fighter aircraft.

One of the most traditional aviation companies, SAAB was founded in Sweden in the early 1930s. Since then, the company with 13,000 employees has been producing weapon systems for the land forces and navy, combat aircraft and military and civil multi-role aircraft, is active in space travel and manufactures information technology and simulators. The development and production of the Gripen combat aircraft, the development of anti-tank guided weapons, are particularly worthy of mention. The Volvo Aero company is also very successful in the aviation industry and is considered a global supplier of engines for all types of aircraft. Volvo engines or engine parts are now found in around 80% of all engines on larger aircraft. For the sake of completeness, it should be mentioned that the flight academy of the Scandinavian airline SAS offers training services to the air force.

Bofors, a company rich in tradition, can look back on centuries of company history and is considered the producer of the most successful anti-aircraft guns of the Second World War. Today, the company is owned by BAE and produces air defence systems, high-precision naval guns and intelligent ammunition for the manufactured products.

Sweden has a long naval tradition with excellent shipbuilding products. The VASA, which sank during the launch, was the exception. However, the sinking of the VASA was a blessing for posterity, because in the VASA museum you can study 18th century shipbuilding on the basis of the preserved ship. Today, Kockums continues the tradition of shipbuilding. The company is owned by the German HDW and develops and produces Gotland-class submarines and Visby-class stealth corvettes at its Malmö and Karlskrona sites.

In addition to the companies mentioned above, there are a number of companies that produce electronics

4.19. Switzerland

History

Celtic Helvetians invaded the area of present-day Switzerland between the Jura and the Alps around 100 BC. They were defeated by Julius Caesar around 58 BC and the area was incorporated into the Roman Empire. Around 15 B.C. the eastern part of what is now Switzerland was also incorporated into the Roman Empire. During the Migration Period, the Alemanni conquered those parts of Switzerland that today comprise the German-speaking area. In the 6th century, the present national territory was incorporated into the Frankish Empire. This resulted in several divisions of the empire, for example Eastern Switzerland became part of Eastern Franconia in 843 and Western Switzerland became part of the German Empire in the 11th century. In the 14th century there was a fight against the Habsburgs, which ended with crushing defeats of the Habsburg army of knights against the federal peasant army. The victory of the Swiss Confederates in the Swabian War in 1499 against Emperor Maximilian I led to their secession from the German Empire, but this was not confirmed until the Peace of Westphalia in 1648. Switzerland's permanent neutrality was recognised at the Congress of Vienna in 1814/15. Since then Switzerland has remained neutral in the wars.

Defence Industry

The technology company RUAG is an internationally active specialist company with a high level of technological expertise in aerospace and aeronautics as well as in the renewal of weapon systems and the manufacture of a wide range of ammunition types. The holding company is based in Berne with production sites in Switzerland, Sweden and Germany. As part of its armaments policy, the owner requires RUAG, among other things, to secure important industrial know-how for the benefit of the Swiss Armed Forces through international cooperation, so that the Armed Forces can fulfil their constitutional mandate thanks to the appropriate material readiness.

The aerospace division is ISO 9001 certified and deals with both military and civil technologies. In the field of military aviation, one of the most important customers is the Swiss Air Force. For the Swiss Air Force, the F-5 Tiger aircraft were modernised, the PC-7 fleet was repaired and the Swiss Air Force's modern F-18 Hornet fighter aircraft underwent a combat upgrade at the Emmen plant. The maintenance of the Swiss Air Force's helicopter fleet is also part of RUAG's business.

The RUAG Electronics division is an extremely successful subsidiary in the development of simulators and training equipment and also deals with tailor-made solutions for C4ISTAR. This covers the areas of Command & Control, Communication, Computers, Intelligent Products, Surveillance and Target Acquisition and Reconnaissance, all research and development fields that will be of particular importance in the future NetWorCentric-Warfare scenario.

Another division, RUAG Land Systems, develops, manufactures and supplies complete weapon systems, including the necessary army logistics, such as spare parts management and sizing, as well as configuration management and the defect reporting system. The activities range from applied research to operational support. This business unit is concerned with both the battle value enhancement of existing weapon systems, such as the battle value enhancement of the M-109 artillery system and the M-113 light tank family, and the development of intelligent solutions for mine launchers, where a 120mm mine launcher has been developed.

The RUAG Components division is regarded as an innovative developer of complex components for intelligent ammunition systems as well as steel and aluminium bullet casings with metal, plastic and steel guide strips.

RUAG Ammotec is an international defence technology competence centre for the development, production, maintenance and sale of ammunition, assemblies and ammunition systems. At the disposal facility for ammunition components and explosives, RUAG Ammunition disposes of ammunition components and residual materials that can no longer be recycled in an environmentally friendly manner. RUAG Ammotec operates production facilities in Thun and Altdorf. Since the acquisition of Dynamit Nobel and similar companies, RUAG has become the largest manufacturer of small calibre ammunition in Europe.

MOWAG (Motorwagenfabrik) AG was founded in 1950 as a private company and has concentrated over the past 50 years on the development and production of special vehicles for the military sector. It is considered one of the leading suppliers of armoured and unarmoured military vehicles in the entire range of weight classes from 7 to 25 tons. Thanks to the quality of its vehicles in terms of high protection value and great mobility, MOWAG has achieved a market position on the world market which is reflected in impressive sales figures. Today, there are about 8,000 Piranha wheeled armoured vehicles, about 500 4x4 Eagle armoured reconnaissance vehicles and about 3,500 Duro transport vehicles in use worldwide.

Since 2003, MOWAG has been part of the General Dynamics European Land Combat Systems Group. General Dynamics is one of the world's leading defence groups and generates its sales with tactical IT systems, land-based and amphibious weapon systems, shipbuilding and business aircraft. Oerlikon Contraves was founded in 1906 as a machine factory and has been involved in the development of anti-aircraft weapon systems since the mid 1930s. The success of Oerlikon guns is primarily based on the period of the Second World War, when Oerlikon-Bührle supplied the Wehrmacht with a large number of guns and ammunition. In 1999 the company was taken over by Rheinmetall DeTec. Oerlikon is currently the world market leader in the development and production of tube-guided antiaircraft weapon systems and has subsidiaries in Switzerland (headquarters), Germany, Italy, Canada, Malaysia and Singapore. Oerlikon currently supplies its products to around 40 countries. The most important and at the same time most successful products are the Skyguard fire control system, Skyshield and ammunition types for anti-aircraft defence.

In 2000, the former SIG Arms Waffenfabrik was spun off from the Swiss industrial company and taken over by two German investors under the new brand name Swissarmss. The high quality standard of SwissArms is based on a strong foundation of experience. As the former SigArms AG, the new company can look back on over 140 years of tradition in weapon design. At the same time, the company has built up a worldwide competence in the course of its history and is present with subsidiaries worldwide. Many armed forces in the world use pistols, submachine guns and automatic rifles from this traditional Swiss weapons manufacturer.

Pilatus Flugzeugwerke was founded in Stans in 1939 and currently employs 1,200 people. The Pilatus Flugzeugwerk develops and produces reliable small aircraft, of which the Pilatus PC-6 (Porter) has been the greatest success to date. Following its presentation at the Paris Air Show, around 410 of the so-called "Jeep of the skies" have been sold worldwide to date. Other business segments of this successful company are training aircraft, such as the PC-7 turboprop training aircraft and aircraft for purely civilian use. The Austrian Armed Forces also use PC-6 and PC-7 aircraft. The company recently presented its new development, a state-of-the-art PC-21 training aircraft, to the public. International experts believe that the new aircraft has excellent market prospects.

In addition to the large corporations, a whole series of smaller Swiss companies have also carved out their niches in the defence industry. These include, for example, Vectronix AG in Heerbrugg SG, which produces night vision devices and supplies them to the US Army, I.L.E.E. AG in Urdorf, which produces target lasers for rifles, and Brügger&Thonet in Thun, which manufactures various weapon accessories.

At the beginning of the 21st century, the Swiss defence industry presents itself as a powerful industrial sector which, through restructuring and mergers with major international groups, is initially securing its ability to survive in a market that continues to shrink. However, the future commercial success of the Swiss defence industry will depend on Switzerland's security and defence policy orientation and the integration of the defence industry into the European network.

4.20. Serbia

History

Serbia's rise to central Balkan power began in the second half of the 12th century. The prerequisite for this was the impotence of the Byzantine Empire, under whose sovereignty Serbia had been under since the 10th century. Under the leadership of the Nemjids, the union of the Serbian tribes began from the Principality of Rasciesa. In 1217, Stephen I was crowned king of independent Serbia. At the end of the 13th century, parts of northern Macedonia and Bosnia were won. The greatest expansion reached Greater Serbia under Stephen IV around 1355, at that time stretching from southern Bosnia, along the Dalmatian and Albanian coasts, Macedonia and central Greece, to Bulgaria. In 1346 Stephen was crowned Tsar of all Serbs and Greeks. Soon after his death, the empire disintegrated into small dominions, some of which became independent. Albania. Macedonia and Bosnia were lost again. From the middle of the 14th century the Ottomans had entered European soil. The Balkan peoples were defeated by the Ottomans in the Battle of the Blackbird Field on 15.6.1389. With this battle trauma of all Balkan peoples, the Turks had succeeded in making the decisive breakthrough in the Balkans. Until 1459 all Serbia was conquered by the Turks. Under the impression of the French Revolution, the Balkan peoples' struggle for freedom began against the Turks and also against Austrians and Hungarians. The Serbs rose in 1804 under the leadership of Karadjordje. Serbia now became an autonomous principality that had to pay tribute to Turkey. In 1878 Serbia finally gained full independence.

Defence Industry

In the territory of today's Serbia, not including Montenegro, the most important armament factories of Yugoslavia, about 43% of the total armament industry, were located at the time of the entire state. These enterprises were all state-owned. After the Balkan wars, if the scarce sources are to be believed, only 11 state-owned enterprises with a workforce of around 15,000 remained, and another drastic measure is still to come, as the industry intends to carry out extensive privatisation, so that only 6 enterprises with around 5,000 employees will remain state-owned. The most important companies are Zastava's arms factory in Kragujevac, which survived the NATO bombing to some extent, and Prvi Partizan. Prvi Partizan in Uzice was founded in 1928 and produces various types of pistols, rifles, about 300 different types of ammunition and machines for ammunition production and is one of the largest ammunition manufacturers in the world. In the aviation industry, the company UTVA, which produces the LASTA 95 training aircraft, should be mentioned.

The Swiss arms industry is diverse, as the two pictures on the next page show: Aircraft, drones, anti-aircraft guns, small arms, assault rifles, vehicles, fire control units, armoured all-purpose vehicles, wheeled armoured vehicles, grenade launchers.

Serbia's defence industry produces armoured vehicles

With SAAB, Sweden has an efficient producer of military equipment, which also manufactures combat aircraft (Gripen).

4.21 Slovakia

History

Before the birth of Christ, the territory of present-day Slovakia was a settlement area of the Celts and later of the Germanic Quades. Around 500 AD Slavic tribes migrated to the area. They founded several empires in the area. From the 11th century Slovakia came under Hungarian rule. Only in 1918, after the collapse of the Danube Monarchy, Slovakia managed to escape Hungarian rule. Slovakia founded Czechoslovakia together with Bohemia and Moravia. From 1939 to 1945 Slovakia became temporarily independent, but had to coordinate its policy with the German Reich. After the Second World War, Czechoslovakia was restored with some territorial separations. From 1969 after the Prague Spring Slovakia became a partial republic. After the collapse of communism at the end of the 1980s, the division of Czechoslovakia by mutual agreement between Czechs and Slovaks was decided. Formal independence has existed since 1.1. 1993 with the foundation of Slovakia. Slovakia is a member of the UNO, the Council of Europe, and since 2004 of NATO and the EU.

Defence Industry

In its best period in the mid-1980s, the arms-producing industry of Czechoslovakia had an armaments industry in what is now the Slovak Republic, employing some 80,000 people and producing armaments worth some US$519 million. After the collapse of the WAPA, the number of employees shrank to 6,000 and, according to a mailing by the Association of Czech Defence Equipment Producers, it currently produces military equipment worth the equivalent of 29 million US dollars in some 40 companies.

After the so-called "velvet divorce" from the Czech Republic, the Slovak Republic was left with only one air carrier, LOT Trencin, which specialises in the maintenance of aircraft from the Soviet era. The company works mainly for the government and for third world countries. The most important business partner became Egypt, which has its L-29 Delfin and L-39 Albatros training aircraft overhauled by LOT. LOT has also modernized new SU-22 fighter aircraft for Angola.

In the field of hard armaments there are Delta Defence, Dubnicky Metalurgicky Kombinat, ZTS-OTS, Kerametal and some ammunition factories. Kerametal is a private company that financed the development of the Aligator family of vehicles. In addition, there are several companies in the electronics industry, which primarily offer modernizations of military equipment in use.

Despite all efforts to keep the Slovak defence industry alive, its future is more than uncertain due to the existing financial difficulties in which the entire Slovak industry finds itself.

4.22. Spain

History

In antiquity there was a colourful mixture of peoples on the Iberian Peninsula. Probably the oldest known population are the Basques. Beside the Basques, Ligurians and the Berber Iberians, who came from North Africa, settled down. Since the 6th century B.C. Celts settled and merged with the Iberians. Even before Christ, Phoenicians and Greeks established trading posts. The Carthaginians conquered Spain in the 3rd century B.C. In the 2nd century B.C. they were expelled by the Romans. With the migration of peoples, Germanic tribes came to Spain. In the 8th century A.D. finally Islamized Berbers came to Spain and established the Umayyad Caliphate, which was to remain in existence until the Reconquista. At the end of the 15th century the expulsion of the Islamic rulers succeeded and a new glorious age began. Spain quickly rose to become a world power, conquered almost all of Central and South America and maintained colonies in Africa and Asia. After the defeat against England in 1588, the slow decline began, which lasted until the independence of the colonies in the first half of the 19th century or until the end of the 19th century. This was followed by a restoration. Spain managed to stay out of the First World War. After internal unrest, the Spanish Civil War broke out, which lasted from 1936 to 1939. Afterwards Franco succeeded in stabilizing Spain and keeping it out of the Second World War. After the death of Franco in 1975, the monarchy was reinstated and Juan Carlos I was crowned king. Spain is a member of the EU and NATO.

Defence Industry

The Spanish defence industry has a wide range of companies that can provide defence equipment for all branches of the armed forces. As a result of an extensive privatization process between 1996-2003, the state-owned defence companies were restructured and reorganized within the SEPI (Sociedad

Estatal de Participaciones Industriales) holding company. The most important defence companies within SEPI are the shipbuilding yard IZAR, the electronics group INDRA SISTEMAS and the aviation company EADSCASA. The importance of the Spanish defence industry is also manifested by the fact that at least IZAR is ranked 56th in a ranking of the 100 most successful defence companies in the world with a turnover in 2003 of around 740 million US dollars from defence business.

The Navantia shipyard is the result of the merger between Asti and Empresa Nacional Bazan and is the second largest shipyard in Europe with large facilities in Ferrol, Cadiz, Cartagena. Navantia offers a wide range of products from the development and production of combat ships and on-board weapon systems to the repair and modernization of older weapon systems. As the shipyard's products have been among the best of their kind for decades, the shipyard's customers include not only the Spanish Navy, to whom it has supplied an aircraft carrier, frigates, submarines, smaller combat ships and also landing ships, but also the Thai Navy, for whom it has built an aircraft carrier, and the navies of Morocco, Angola, Argentina, Chile, Mexico, Peru and Portugal, for whom it has built frigates, corvettes, patrol boats and submarines.

Spain can also boast of efficient companies in the electronics sector. The electronics company Indra Sistemas is the leading Spanish electronics company which cooperates closely with the French electronics giant Thales, formerly Thomson-CSF, and the Spanish armed forces in the field of military electronics and offers military-specific IT solutions.

The aircraft manufacturer Airbus Defence and Space is part of the European Airbus consortium. Since its foundation in 1923, the Spanish aviation company Construcciones Aeronauticas S.A. (CASA) has constantly developed a technological and productive capacity that today enables it to compete in the international aerospace market for design, manufacturing and maintenance contracts. The Getafe production facility designs and manufactures transport aircraft, but it has also carried out final assembly of the Eurofighter combat aircraft destined for Spain and modernized the P-3 Orion maritime surveillance aircraft, among others.

The largest defence contractor for the production of infantry weapons, ammunition and land systems is General Dynamics Land Systems (GDELS) Santa Bárbara Sistemas. The company can look back on a long tradition, as the first production facility for weapons was founded in Seville as early as 1540. The company belonged to SEPI Holding; in the course of restructuring, the company was taken over by the US-American armaments group General Dynamics Corporation and since then has been one of the most important European suppliers of combat vehicles, special and amphibious vehicles, weapons and weapon systems and ammunition. One of the most important combat vehicles is the Pizarro main battle tank, better known internationally as ASCOD, which is also produced by the Austrian subsidiary of General Dynamics Steyr Daimler Puch Spezialfahrzeuge as Ulan for the Austrian Army. In addition to the Pizarro, the company also builds smaller combat vehicles and, under license, the German Leopard 2 main battle tank. Artillery weapon systems and artillery ammunition are another line of production at Santa Barbara. In the field of small arms and machine guns, Santa Barbara produces both in-house developments and licensed products from the German weapons manufacturer Heckler&Koch. For all weapon systems developed and produced by Santa Barbara, a wide range of ammunition types are manufactured at the production facility in Palencia.

A major manufacturer of off-road vehicles is URO, Vehículos Especiales, S.A (UROVESA).

4.23. Czech Republic

History

The territory of the present-day Czech Republic has been the scene of ethnic, political, economic, religious, national and social conflicts for more than 2000 years. The first states were formed by Germanic states. The Moravian region was ruled by the Samo people in the 7th century. In the 9th century the Great Moravian Empire was formed. After the disintegration of this empire a state was formed by the Premyslids. The empire ended with the defeat of King Ottokar against Rudolf of Habsburg in 1278 in the Battle of Jedenspeigen. After the Premyslides the Luxembourgers ruled. Among the Luxembourgers there were the Hussite Wars between 1419 and 1436. On 1526 the Habsburgs ruled over Bohemia and Moravia. Bohemia and Moravia were part of the Habsburg Empire during this time. Only on 28.10.1918 Czechoslovakia became a republic. 1938 Czechoslovakia was forced to cede the Sudetenland to the German Reich. During the Second World War Bohemia and Moravia were occupied by German troops. At the end of the Second World War the country was liberated by the Soviet Union and increasingly came under the influence of the Soviet Union. In 1948 the regime loyal to the Soviet Union was consolidated. In 1968 there was an uprising, which was crushed by the invasion of troops from the member states of the Warsaw Pact. After the end of communism, the country was divided by mutual agreement into a Czech Republic and a Slovakia. Formal independence has existed since 1.1.1993. The Czech Republic is a member of the UN, the Council of Europe and since 1999 of NATO and in 2004 of the EU.

After the foundation of Czechoslovakia, the young republic took over the former Austro-Hungarian heavy industry in Bohemia, Moravia and Slovakia. In addition, Czechoslovakia benefited from the fact that one of the world's largest armament producers, the Skoda Works in Pilsen, was now located on its territory. Another important industrial group was Ceskomoravska Kolben Danek (CKD), which was created by the merger of the First Bohemian-Moravian Engineering Works in Prague with Kolben&Co and Danek. The CKD concern employed about 12,000 workers in 1927; between 1946 and 1990 the number increased to 50,000. The CKD concern produced Praga automobiles. Another important building block for the national defence industry was Tatra, a car and truck manufacturer. The new state was particularly interested in building up powerful armed forces. This required tanks, aircraft and other heavy weapons. The Skoda Works had great experience in the production of cannons and howitzers. In the mid-1930s new guns were developed and produced, such as the 149 mm M 25 and M 37 field howitzers and the 105 mm M 35 field howitzers. Small arms submachine guns and machine guns were built in the Ceskoslovenska Zbrojovka Brno and Ceska Zbrojovka Strakonice weapon factories built after the First World War, and the armed forces were equipped with them.

Like many other nations, Czechoslovakia gained experience in tank construction by importing French Renault FT and British Carden-Loyd. At the beginning of the 1920s Josef Vollmer, the former chief designer of the motor vehicles department in the German War Ministry, came to Prague. He went to Skoda and started to develop a light tank. The development of small tanks was pushed by Skoda/Tatra and CKD/Praga. Skoda/Tatra developed the LT-35, a technically very advanced tank. The combat weight was about 10 tons. The crew consisted of four men, the armament was a 3.7 cm Skoda L/40 board cannon and two 7.92 mm machine guns. The development of

CKD/Praga, the tank TNHB/LT -34 (Lekkhy Tank/Light tank introduction year 1934), was also exported to half a dozen countries. The LT-34 had a combat weight of 7.6 tons and a crew of 4 men. The armament consisted of a 3.7 cm Skoda L/40 board cannon and two 7.92 mm machine guns. Due to the great export success the TNHB was used as a basis for the further development to the TNHP/LT-38. The LT-38 weighed 9.7 tons and had a crew of 4 men. The armament consisted of a 3.7 cm Skoda L/47.8 board cannon and two 7.92 mm machine guns. After the invasion of the German Wehrmacht and the occupation the LT-38 was continued to be built and used in the German Army with great success. A total of 1,590 units were built. The chassis of the LT-38 tank also served as the basis for a tank family that included a FLAK tank, a self-propelled gunner and the "Hetzer" jadg tank.

The Austro-Hungarian Danube Monarchy had no aviation industry worth mentioning. The young republic therefore had to build up its own aviation industry, which performed and still performs remarkable services for the small country. AVIA Flugzeugwerke was founded just one year after gaining independence. Among the most important aircraft productions were the fighter planes Av-135 and Av-534. 445 of the Av-534 were produced. Avia ceased aircraft production in 1960, but continued producing aircraft engines until 1988. Two other companies, Aero Vodochody and Letov were also founded in 1919 and also built aircraft for the Czech armed forces. Aero Vodochody still produces jet-propelled training and light ground combat aircraft today. The fourth company, Moravan Otrokovice, was founded in 1934 and built aircraft with the type designation Zlin.

In its best period around 1987, the armaments manufacturing industry of Czechoslovakia had an armaments industry in what is now the Czech Republic, employing around 100,000 people and producing armaments worth around 5.3 billion US dollars. After the end of the Cold War, the number of employees shrank to 17,000 and, according to a statement by the Association of Czech Defence Equipment Producers, around 130 companies currently produce military equipment worth 175 million US dollars, of which around 100 million US dollars is accounted for by the aviation industry. The most important producers of armaments are engaged in aeronautical engineering, armored vehicles, modernization of armored vehicles, especially battle tanks, communication equipment and electronics, protection systems against nuclear, biological and chemical (NBC) weapons systems, as well as small arms and ammunition.

The armoured vehicle production sector is dominated by state-owned companies specialising in the modernisation of combat vehicles. One of the largest companies in this sector is VOP Cz Another larger company is the military vehicle manufacturer Tatra.

The electronics and communications technology sector is dominated by the companies MESIT and Tesla, which have long experience in the production of military equipment.

The small arms and light weapons sector is dominated by the Ceska Zbrojovka (CZUB) company. The produced weapons are exported to the whole world. Since 1997 there is also a branch plant of CZ in the USA. The ammunition sector is dominated by the companies Sellier&Bellot and ZVI.

The production area of protective equipment against NBC threats is dominated by the companies Ortitest and Gumarny Zubri. While Ortitest manufactures test and decontamination equipment, Gumarny Zubri produces NBC protective masks and clothing.

The Spanish defence industry is an important branch of industry in the country and also produces aircraft. This is a transport plane.

Spanish shipyards also produce large combat ships. Here in the picture an aircraft carrier for the Spanish Navy

The Czech defence industry is able to produce jet trainers

Czech Republic is an important supplier of NBC protection equipment

The Slovakian arms industry produces, among other things, artillery guns

4.24. Turkey

History

In the 11th century Turkish tribes from Central Asia immigrated to the area of today's Turkey. Around 1300, the Turk called on Osman I to wage a religious war against the Byzantines and he drove Byzantium out of West Asia Minor. Through his conquests he laid the foundation for the Ottoman Empire. Already in the middle of the 14th century the Turks invaded Europe and defeated the army of the united Balkan princes on the Blackbird Field on 15.6.1389. In 1453 the Turks finally succeeded in conquering Byzantium, after which Serbia, Bosnia, Albania and other parts of the Balkans were conquered. In 1529 Vienna was finally besieged unsuccessfully. In the 16th century Turkey gained further territories in Mesopotamia and North Africa. The decline began in 1600 and reached its temporary peak with the defeat of the Turkish army in 1683 in front of Vienna. Subsequently, the imperial army succeeded in reconquering large areas of the Balkans. Russia became the Turks' main enemy. In several wars Russia expanded its empire. In 1827 the

Greeks succeeded in gaining independence by defeating the Turkish-Egyptian fleet with a victory of the united fleets of England, France and Russia. In the Crimean War of 1853, Turkey succeeded in limiting Russia's supremacy in alliance with England and France. From 1883 the Turkish army was trained by German soldiers. Turkey entered the First World War on the side of the Central Powers. In 1917 British troops advanced into Palestine and Syria. After the armistice in 1918 Greek troops occupied Smyrna. After the defeat, Turkey lost large areas in the peace of Sevres in 1920 and was restricted to Anatolia. Smyrna was awarded to the Greeks, this led to the war against Greece. Now the general Mustafa Kemal (later Atatürk) placed himself at the head of the national movement. Kemal expelled the Greeks from Smyrna. In 1923 Turkey became a republic and a catch-up process began. Kemal undertook fundamental reforms for modernization. There was a separation of state and religion. The Latin script was introduced. During the Second World War, Turkey remained neutral. After the war, Turkey joined NATO.

Defence Industry

In Turkey, both state-owned and privately owned companies in the defence industry develop and produce light weapons, ammunition, armoured combat vehicles, naval vessels and combat aircraft. The defence industry companies are in close proximity to the Ministry of Defence and the armed forces. The most important company for ammunition production is Makina Ve Kimya Endüstrisi Kurumu (MKEK), which was founded in the 15th century and manufactures small arms, artillery ammunition, fuses, hand grenades, aerial bombs, weapon containers for aircraft in 12 production plants.

For the development and production of land systems, the Roketsan company specializes in the production of pistols, grenade rifles, rocket launcher systems, anti-aircraft missiles for terrestrial and sea-based operations, and even battle tanks. For example, the company produces the French Leclerc main battle tank under license for the Turkish armed forces. The company is also involved in the development of unmanned aerial vehicles and precision guided weapons for combat aircraft.

The company Otokar, founded in 1963, also specializes in the development and production of land systems. In addition to licensed products such as the Landrover, it also produces light armoured wheeled vehicles such as the Cobra, Otokar APC and the Akrep. Another company that produces combat vehicles is the armaments company FNSS, which emerged from a joint venture between the US armaments company United Defense and the Turkish company Nurol Holding and has been producing main battle tanks and infantry fighting vehicles since 1998.

The most important company in military aviation is Turkish Aerospace, founded in 1984, which in the past assembled US F-16 fighter planes for the Turkish and Egyptian Air Force under license at its production facility in Ankara. Turkish Aerospace also produces aircraft of other makes under license, for example the CN-235 model from CASA and the Cougar AS-532 helicopter. In addition, the company also works for the Turkish Air Force as a general contractor for the modernization of the air fleet. Another company in the aviation industry is TEI, a Turkish General Electric subsidiary specializing in the production of aircraft engines.

Turkey also has an extremely successful electronics company, Aselsan Elektronik Sanayu ve Ticaret A.S., which has been developing and manufacturing a wide range of telecommunications equipment for the armed forces, radar systems, laser systems for targeting and systems for Command&Control at its Ankara production facility since 1975, and night vision equipment, laser rangefinders and thermal imaging equipment at its Akyurt plant.

Turkey also has a number of efficient shipyards in the naval shipbuilding sector, which produce combat and support ships and submarines for the navy and also carry out the necessary repairs and modernisation. The most important naval shipbuilding yards are in Gölcük, southeast of Istanbul, Tashizak and Pendik.

When presenting the Turkish defence industry, Aspilsan, which develops and produces batteries for the Turkish land, sea and air forces and is also a successful exporter of its products, should not go unmentioned.

4.25. Ukraine

History

Ukraine is the third largest successor state of the former Soviet Union. When the Ukrainian Parliament proclaimed Ukraine's independence in August 1991 and this declaration was confirmed by the population in a referendum, it effectively meant the end of the Soviet Union. Although these events took place more than a decade and a half ago, the process of transformation of the political, economic and social system in Ukraine is far from complete. Ukraine's foreign and security policy is largely determined by its geopolitical position between the West and the Central Eastern European states on the one hand and the Russian Federation on the other. Factors influencing foreign policy are, in particular, the strong Russian minority and the regional division of Ukraine with regard to the desired orientation of foreign and security policy. Nevertheless, within a few years, Ukraine has succeeded in securing its existence as a state through a series of international treaties, thereby establishing external legal security. Basic treaties have been concluded with the neighbouring states Hungary, Poland, Moldova, Slovakia, Romania, the Russian Federation and Belarus. These treaties were supplemented by special border regulation agreements.

Defence Industry

In the past, Ukraine was primarily a supplier of technologically complex components and electronic parts, most of which were needed for integration into armament systems at final assembly plants in Russia. As a result, a large number of research institutes were active in the Ukraine. After the dissolution of the USSR there was also an interruption of the close economic cooperation, so that the Ukraine is increasingly trying to catch up with Western companies.

The Ukraine has the traditional tank factory in Charkov, which produced tanks as early as the 1930s. As orders are lacking, production is now only at around 30% capacity. At peak times, the plant was able to produce about 800 battle tanks per year. Today the aerospace industry of Ukraine comprises about 20 companies. Among them are development and research institutes. The largest manufacturers of aerospace products are the Antonov aircraft factory in Kiev, founded in 1946, and the Motor Sich maintenance company.

Shipbuilding also has a long tradition in Ukraine. As early as 1788 a shipyard was built in Nikolayev on the Black Sea, which became one of the most important shipyards for the construction of naval battleships during the Soviet era. Nowadays, due to lack of money, naval shipbuilding has largely ceased, so that the shipyard had to switch to civilian products. Only the final outfitting of already completed combat ships and the modernisation of older units are carried out in Nikolayev near Chernomorsky Shipbuilding Yard.

4.26. Hungary

History

Hungary celebrated the millennium of its founding in 2000. The first Hungarian king, Stephen the Saint, was crowned in the year 1000. After various royal houses, the Hungarian throne was occupied by the Habsburg dynasty from the 16th century. For centuries, Hungarian politics was characterized by the opposition of the pro and contra Habsburg parties. With the division of Hungary after 1918, a national territory with an ethnically almost homogeneous population was created. The period after the First World War can be divided into the phase of the semi-democratic parliamentary system of government between 1920 and 1944, the phase of communist one-party rule from 1949 to 1989, and the phase of a democratic republic with a parliamentary system of government since 1990. The democratic change in Eastern Europe led to the withdrawal of Soviet troops in Hungary. After regaining sovereignty, Hungary found itself in a situation of renewed and old-fashioned insecurity. Tensions with neighbouring states, which were a consequence of the Trianon peace treaty of 1921, broke out again. In order to overcome the new insecurity, Hungary increasingly oriented itself towards Western Europe and strove for integration into its institutions. The foreign and security policy focus of every government was European policy.

Defence Industry

The defence industry in Hungary reached its peak in 1988, when about 40 companies with 30,000 employees generated a turnover of about 370 million US dollars. The export share was about 75%. After the end of the Cold War, there was a decline in the armaments industry, making the Hungarian armaments industry one of the smallest in Central and Eastern Europe.

Hungary's defence industry specialises in the repair and modernisation of aircraft and combat vehicles, electronic equipment, radar, simulators, telecommunications equipment, ammunition, small arms and protective equipment.

The armoured vehicle sector has been small in Hungary up to now and currently only Currus is modernising the old BTR-80 and BMP-1 infantry fighting vehicles.

The most important company for the production of vehicles is Raba-Werke. In the course of the modernisation of the Hungarian armed forces, Raba, which also cooperates with the German MAN, can expect a major order within the next 15 years.

The Hungarian electronics industry and software development has suffered a steady decline in importance over the last decade, yet a handful of companies produce electronic equipment.
Two particularly successful companies are active in the production of protective clothing against the effects of nuclear, biological and chemical weapons. Respirator and CBRN Magyarország deliver their products all over the world.

Despite all efforts to keep the Hungarian defence industry alive, its future is more than uncertain due to the existing financial difficulties in which the majority of companies find themselves, and they must strive to ensure that they remain in the market by mergers and joint ventures with potential foreign companies.

Hungary produced an armoured infantry fighting vehicle in WAPA, which was widely used

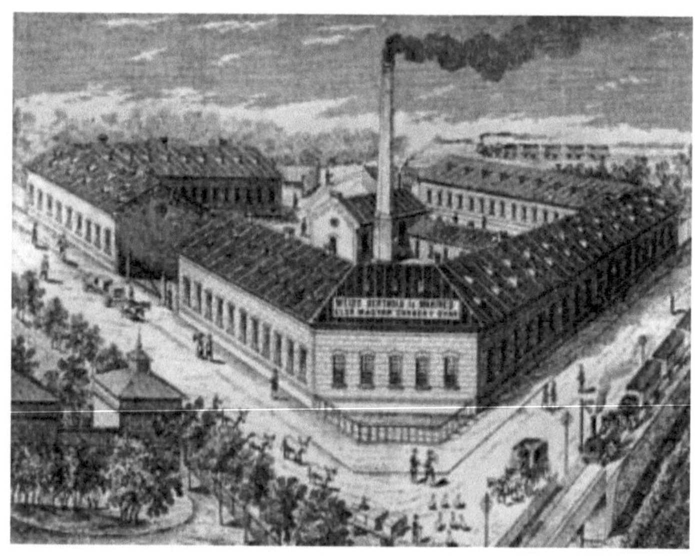

Already in the Habsburg Monarchy, armaments were produced on Hungarian soil, for example by Weiss

Antonov is an important company in the defence industry, which supplies transport aircraft

The arms industry of Belarus is small, but it produces heavy-duty transport systems

Turkey's defence industry is diverse and customer-oriented.

4.27. Belarus

History

The Republic of Belarus was part of Soviet territory until the dissolution of the USSR at the end of 1991. The current political system in Belarus can be described as an authoritarian dictatorship with certain tendencies towards totalitarian structures. In the first years of independence, the country pursued a course of foreign policy neutrality, the primary goal being to break away from Russia and simultaneously open up to the West. Since President Lukashenko took office in 1994, Belarus has been endeavouring to achieve the greatest possible reintegration with the Russian Federation. At the same time, relations with the West have deteriorated.

Defence Industry

After the dissolution of the Soviet Union, the military-industrial complex in Belarus also suffered heavy losses due to lack of demand. The most important armaments plant is the Minotor plant located in Minsk, which modernizes armored vehicles. The Belarusian capital is also home to Minsk Fahrzeugwerke, which offers a wide range of all-terrain trucks. There are also companies that modernize and maintain aircraft.

Printed by Books on Demand GmbH, Norderstedt / Germany